VISUAL ANATOMY

Limbs and Back

VISUAL

Sydney M. Friedman
M.D., Ph.D., F.R.S.C.

PROFESSOR AND HEAD, DEPARTMENT OF ANATOMY
THE UNIVERSITY OF BRITISH COLUMBIA
VANCOUVER, CANADA

VISUAL ANATOMY

VOLUME I: HEAD AND NECK
VOLUME II: THORAX AND ABDOMEN
VOLUME III: LIMBS AND BACK

ANATOMY

VOLUME THREE

Limbs and Back

WITH 87 ILLUSTRATIONS

Medical Department
HARPER & ROW, PUBLISHERS
Hagerstown, Maryland
New York, Evanston, San Francisco, London

VISUAL ANATOMY: VOLUME III. LIMBS AND BACK

 Printed in the United States of America. For information address Harper & Row, Publishers, Inc., 2350 Virginia Avenue, Hagerstown, Maryland 21740.

First Edition

STANDARD BOOK NUMBER: 06-140835-2

LIBRARY OF CONGRESS CATALOG CARD NUMBER: 70-133138

For
Constance Livingstone Friedman

CONTENTS

THE FOREARM AND HAND

THE HAND

PREFACE

This VISUAL ANATOMY series was conceived at a time when standard courses in gross anatomy were ponderous with detail. It is being completed at a time when it is commonplace for students to feel bereft of adequate anatomical information. These wild swings in curricular fashion emphasize the importance of providing middle of the road treatment of a subject so fundamental and relevant to the professional student's competence. I do not expect that this requirement will disappear or even change in any radical way, because an understanding of the structure of the human body provides the basic key to an international medical language. Only evolution can change the body, and the vocabulary of medicine grows more, not less, precise and uniform.

It is unreasonable to expect a student to learn so rich a vocabulary in a new tongue with any less application than he would devote to any other second language. There are many variations in the methods of study appropriate to the subject—none free from the pain of effort. The subject by its very nature is a visual one, however, and this offers a direct path of entry for information. Remembering the arrangement of structures in the axilla is much akin to remembering the appearance of a patient with Graves' disease—both impress a visual memory and are recalled by a mental image. It is this that makes the use of dissected material so essential in learning anatomy, for although the approach of necessity requires the body to be taken apart, the aim is to demonstrate visually how it is put together. VISUAL ANATOMY is not constrained by the order of dissection, so it builds a series of images directly.

The preparation of this final volume has been more challenging than I expected. Although the first two volumes required extensive reworking, they had at least been thought through once before. This third and last volume of the series, however, had never been more than an idea; now it had to be matched in style, form, and brevity with the others. I have enjoyed doing it, however, and hope that it will prove interesting and useful to others as well.

Although coverage of the back was included in Volume II, Thorax and Abdomen, it is equally appropriate to a volume on limbs. Therefore we have repeated this section of four figures and text at the end of this volume.

I am very grateful to my colleague, Dr. Kenneth R. Donnelly, who has once again been unstinting in his careful devotion to accuracy and clinical relevance; any inaccuracies are entirely mine. Dr. Michael Piper has helped greatly by reading and criticizing the manuscript from the standpoint of the surgeon in training. To all of my past students at McGill University and those here at The University of British Columbia I owe much for the encouragement that has led at last to the completion of VISUAL ANATOMY.

It is customary to thank one's wife for forbearance. Now, at the end of the series, I must acknowledge that my wife and colleague, Constance Friedman, has furnished much of the drive and initiative so necessary for this task. She alone made it possible for me to write and illustrate these books without sacrificing the active life of our research laboratory.

Vancouver

The Suspension of the Upper Limb

The Framework:
I. The Anterior Aspect of the Scapula

The **scapula** is a flat, triangular bone placed obliquely against the back of the chest wall. It consists of a thin body which widens at its lateral angle to carry the articular surface, the *glenoid cavity.* The shallow dishing of the glenoid cavity provides a bearing surface for about half the sphere of the humerus at any one time. It is deepened by a rim or lip of fibrocartilage, the *glenoid labrum,* but even so the scapular socket is shallow. The body is little more than an intramuscular plate of bone that provides attachment for the more or less equalized pull of the muscles occupying its anterior and posterior aspects respectively. The bone is often thin enough to be fully translucent. The vertebral or medial border of the scapula is flattened anteriorly to provide attachment for Serratus Anterior. The axillary or lateral margin is thick and robust to reinforce the bone against the deforming stresses of the muscle pulls.

Protruding upward, forward, and laterally from the superior border of the bone is the knucklelike *coracoid process.* Adjacent to the base of the coracoid the superior border presents a well-marked notch, the *scapular notch.* This is bridged by the *transverse scapular (suprascapular) ligament;* when the ligament is ossified, as is often the case, the notch is converted to a foramen.

A large spine projects backward from about the middle of the scapula. This *spine of the scapula* continues laterally as a cantilevered flat surface overhanging the shoulder joint. This overhang is termed the *acromion.*

Much of the scapula can be discerned in the living subject. In particular, the medial border can be palpated from behind as can the spine and acromion throughout their length. Anteriorly, the protruding tip of the coracoid process is readily felt immediately below the lateral third of the clavicle. These points form useful landmarks in the examination of the living subject.

A synovial capsule extends laterally from the scapula to embrace the head of the humerus. The capsule begins at the peripheral margin of the glenoid labrum and widens rapidly to accept the head of the humerus. It is particularly large and roomy inferiorly to allow for the extensive range of movement at the scapulohumeral joint. The tendon of the long head of Biceps arises from a roughness, the *supraglenoid tubercle,* just above the glenoid cavity, and passes laterally across the top of the humerus encased within a fold of the synovial membrane. The tendon thus appears to lie within the shoulder joint although in fact it is excluded from the cavity by synovial membrane.

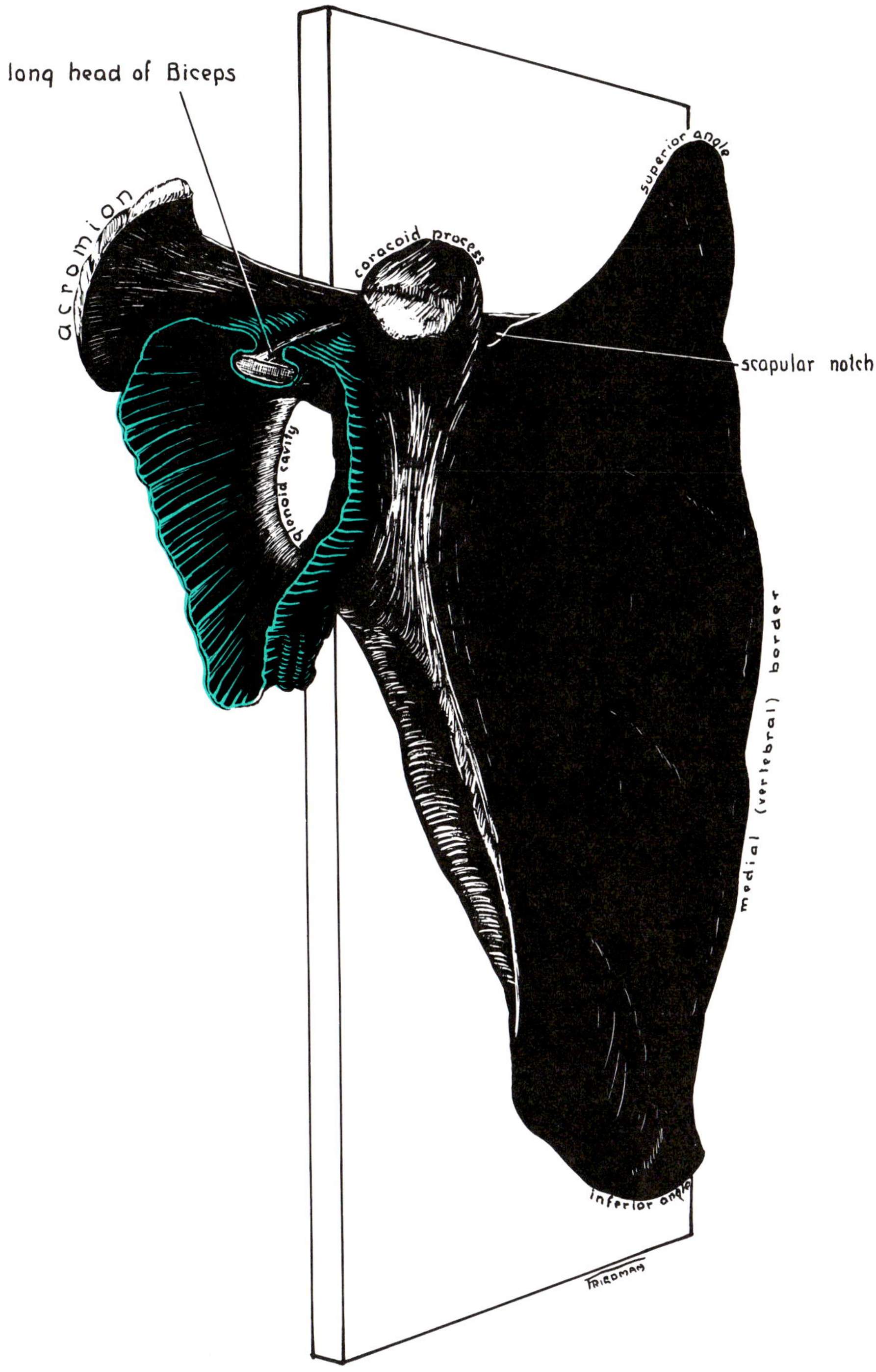
long head of Biceps
acromion
coracoid process
superior angle
scapular notch
glenoid cavity
medial (vertebral) border
inferior angle
FRIEDMAN

The Framework:
II. The Upper Half of the Humerus

The elongated shaft of the humerus expands at its upper end to form the *greater* and *lesser tubercles* and the hemispheric *head*. The function of the tubercles is to support the insertion of the short muscles designed to hold the head in close apposition to the glenoid cavity; these muscles run behind, above, and in front of the articulating surfaces. The continuity of these muscle groups is interrupted anteriorly by the *intertubercular groove,* which lodges the tendon of the long head of Biceps. Most of the muscles involved in keeping the head close to the glenoid fossa insert into the greater tubercle, and only one, Subscapularis, inserts into the lesser tubercle on the medial side of the intertubercular groove.

From the roughened area below the glenoid cavity, the *infraglenoid tubercle,* the tendon of the long head of Triceps arises and passes downward to join with the lateral and medial heads of this muscle about halfway down the arm. While the tendon of origin of the long head of Triceps provides no support to the inferior aspect of the joint when the arm is hanging freely at the side, it rises as the arm is abducted, and comes to lie directly below the joint. Dislocations of the head of the humerus from the glenoid cavity tend to occur when the arm is in the outstretched position and the head of the humerus is driven downward. In this position the Triceps tendon is directly horizontal across the undersurface of the joint, so that when the head is prized out it can move only in front of or behind this tendon. In order to reduce the dislocation successfully it is important to recognize which way the head has gone.

Although the tendon of the long head of Biceps overlies the joint directly and then passes rather abruptly downward in the intertubercular groove, it does not seem to provide any crucial support for the top of the joint.

Since the long head of Triceps runs downward, at first parallel to the shaft of the humerus then gradually converging to meet it, a space remains between bone and muscle. This space is utilized by structures that pass from the front to the back of the humerus.

The acromion overhangs the head of the humerus and is separated from it by an interval that is almost large enough to admit the tip of the finger. This overhang is readily palpated in a living subject and is a little medial to the lateral extremity of the shoulder.

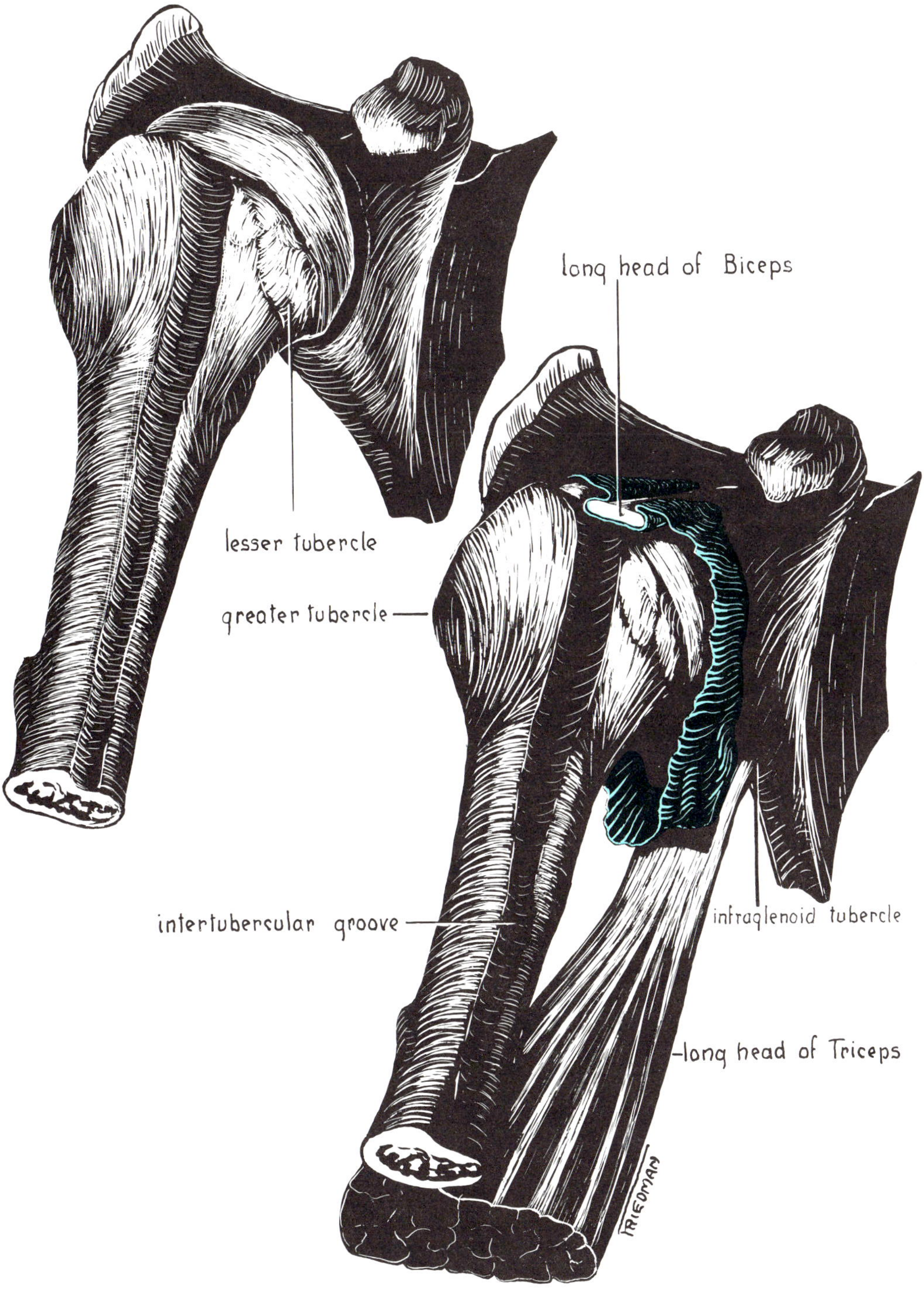
long head of Biceps
lesser tubercle
greater tubercle
intertubercular groove
infraglenoid tubercle
long head of Triceps
FRIEDMAN

The Shoulder Joint:
I. The Synovial Membrane

The *acromion* is connected to the protruding coracoid process by a stout ligament, the *coraco-acromial ligament.* Bone and ligament together complete a *coraco-acromial arch,* which overlies the shoulder joint. As the arm is abducted, the greater tubercle of the humerus comes into apposition with the sharp lateral margin of the ligament and further movement is interrupted. Simple abduction of the arm can thus not proceed much above the horizontal. Moderate external rotation of the humerus, however, frees it from its impingement on the coraco-acromial arch, and the arm can then be further abducted so that the hand can be placed behind the head.

The **synovial membrane,** which was seen in previous illustrations to attach around the margin of the glenoid cavity medially, attaches laterally to the margins of the articular surface of the humerus. A large redundant sac extends inferiorly for about half an inch or more below the head of the humerus. This synovial expansion lies adjacent to the medial aspect of the *surgical neck of the humerus.*

Two extensions of the synovial membrane are always present. One passes forward and medially via a thin neck to form an expanded bursa which underlies the tendon of the Subscapularis muscle. The other forms a synovial sheath or bursa for the Biceps tendon as it lies in the intertubercular groove. This sheath extends down for at least 3 cm on this part of the tendon.

A stout ligament running between the greater and lesser tubercles of the humerus holds the tendon of Biceps in place. This *transverse humeral ligament* is continuous with the fibrous capsule of the shoulder joint.

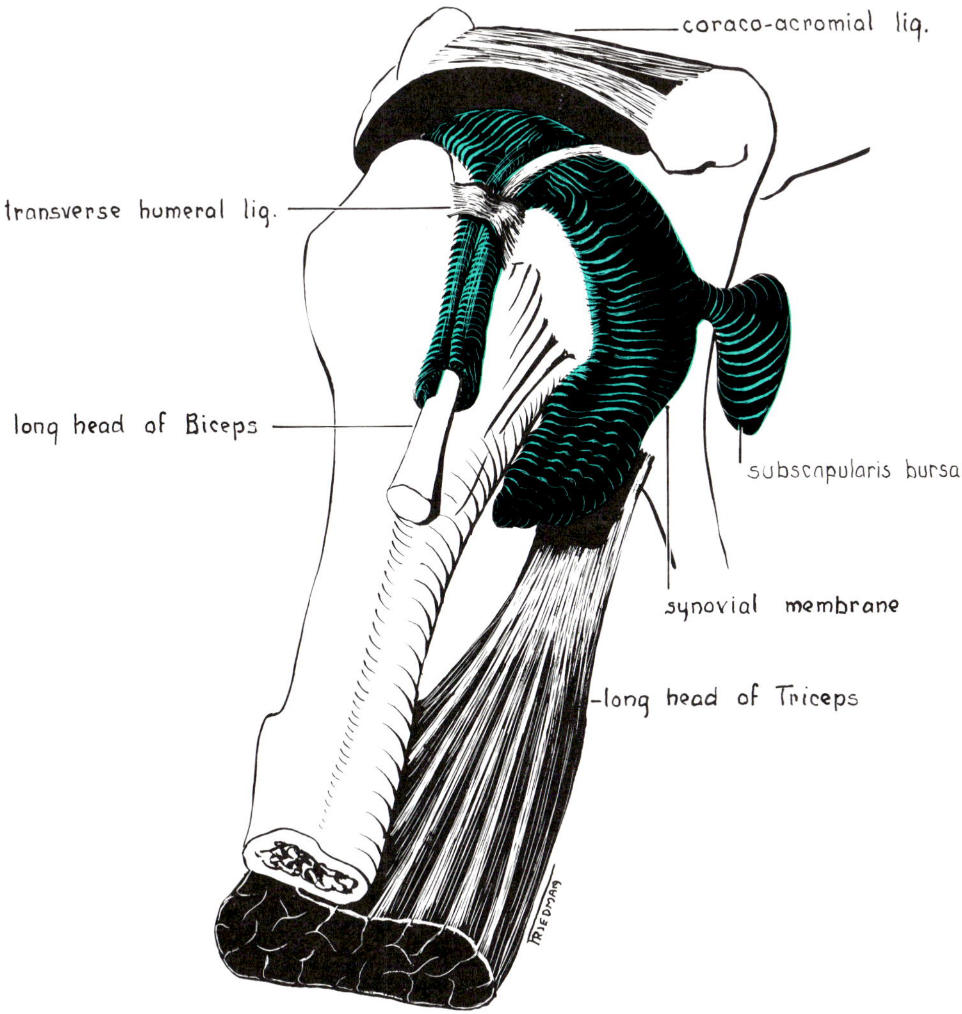
coraco-acromial lig.
transverse humeral lig.
long head of Biceps
subscapularis bursa
synovial membrane
long head of Triceps
FRIEDMAN

The Shoulder Joint: II. The Fibrous Capsule

Overlying and enclosing the synovial membrane is the stout fibrous enclosure for the shoulder joint. This **capsule,** strong in all directions, is very loose; thus, when all muscle attachments are severed it will permit the humerus to be moved away from the glenoid cavity by a centimeter or two. This redundancy is necessary to permit the shoulder joint its full range of movement, which otherwise would be inhibited by inextensible ligaments.

As is usual for all such fibrous capsules, parts of it are identifiable as ligaments. These ligaments are thickenings to provide extra strength in specific directions. Most commonly identified are the anteriorly placed *glenohumeral ligaments,* which are usually subdivided into superior, middle, and inferior groups. A stout *coraco-humeral ligament* running from the base of the coracoid process over the top of the joint is also frequently mentioned. The *transverse humeral ligament* is also essentially a part of the fibrous capsule. The connection between the subscapular bursa and the interior of the synovial capsule occurs between the bands of the glenohumeral ligaments. The tendon of the long head of Triceps arises immediately below the fibrous capsule. It rises to the horizontal position and lies under the joint when the arm is abducted.

The attachment of the fibrous capsule at the scapula is similar to that of the synovial membrane, which it encloses, since it too surrounds the articular margin. At the humerus, however, whereas the synovial membrane follows the articular margin of the head, the fibrous capsule diverges inferiorly to extend down 2 cm or so below the inferior margin of the head. For practical purposes, its lower extent defines the position of the *surgical neck of the humerus.* Fractures of the surgical neck are common; occurring as they do at the junction of the shaft with the expanded upper end of the humerus, they tend to be extracapsular, that is, outside the shoulder joint. In healing, however, they may well involve the joint capsule and the structures related to it, as we shall see.

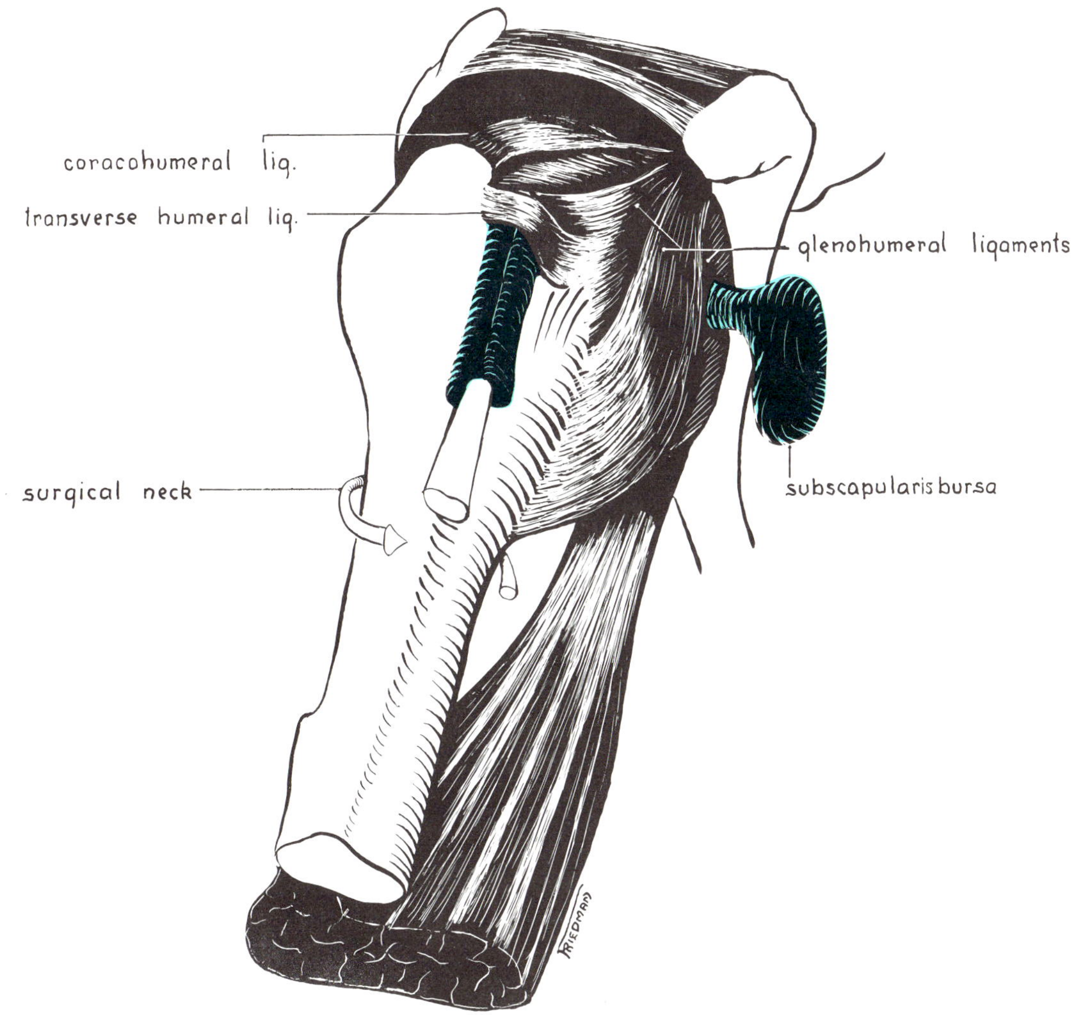
coracohumeral lig.
transverse humeral lig.
glenohumeral ligaments
surgical neck
subscapularis bursa
FRIEDMAN

Muscular Union of the Scapula and Humerus: I. Anterior Aspect

The most prominent muscle anterior to the glenohumeral articulation is **Subscapularis.** This fleshy muscle arises from the medial two thirds of the costal surface of the scapula. It converges to a flat tendon which passes directly across the front of the joint, separated from it by the underlying bursa.

Teres Major may be considered as a downward extension of Subscapularis that finds no room for attachment on the front of the scapula and creeps behind to take origin from an oval, flat area just above the inferior angle of the bone. From here it runs laterally, flattens to a tendon, and inserts into the *crest of the lesser tubercle* (*medial lip of the intertubercular groove*). The lateral half of Teres Major blends with the tendon of **Latissimus Dorsi** which, from a much larger origin in the small of the back, winds upward in front of Teres Major to insert with it. The fibers of Latissimus Dorsi extend right into the floor of the intertubercular groove. The anterior surfaces of Subscapularis, Teres Major, and Latissimus Dorsi together form a smooth musculotendinous posterior wall for the axilla.

The lower border of Subscapularis and the upper border of the Latissimus Dorsi-Teres Major complex separate at their insertion on the humerus and form a triangular interval. The downwardly running fibers of the **long head of Triceps** in the depths of this interval subdivide it into two apertures which link the front of the scapula with the regions behind. One, the *quadrangular space,* lies between the long head of Triceps medially and the surgical neck of the humerus laterally. The other, the *triangular space,* lies medial to the long head of Triceps.

The muscular posterior wall of the axilla (armpit), which we have here defined, is innervated by the posterior cord of the brachial plexus and nourished by the subscapular artery. The **posterior cord of the brachial plexus** runs downward and laterally across Subscapularis where it divides into its terminal branches—the axillary (circumflex) nerve and the radial nerve, its direct continuation. Three nerves arise directly from the posterior cord of the brachial plexus while it lies on Subscapularis. Of these, the *upper* and *lower subscapular nerves* supply the upper and lower fibers of Subscapularis respectively, the lower continuing into Teres Major as well. Between them is the *thoracodorsal nerve,* which runs down to sink into the anterior aspect of Latissimus Dorsi together with terminal branches of the subscapular artery.

The *axillary nerve* passes backward through the quadrangular space together with the *posterior circumflex humeral artery.* Both wind around the surgical neck of the humerus and disappear into the substance of the Deltoid muscle. The fibers contained within the axillary nerve originate mainly from the C5, C6 segments, a point of some importance in localizing nerve lesions.

The **subscapular artery** supplies blood to the muscles of the posterior wall of the axilla; one large branch, the *circumflex scapular artery,* runs backward through the triangular space to supply the muscles of the back of the scapula.

Supraspinatus reinforces the top of the joint (Fig. 7).

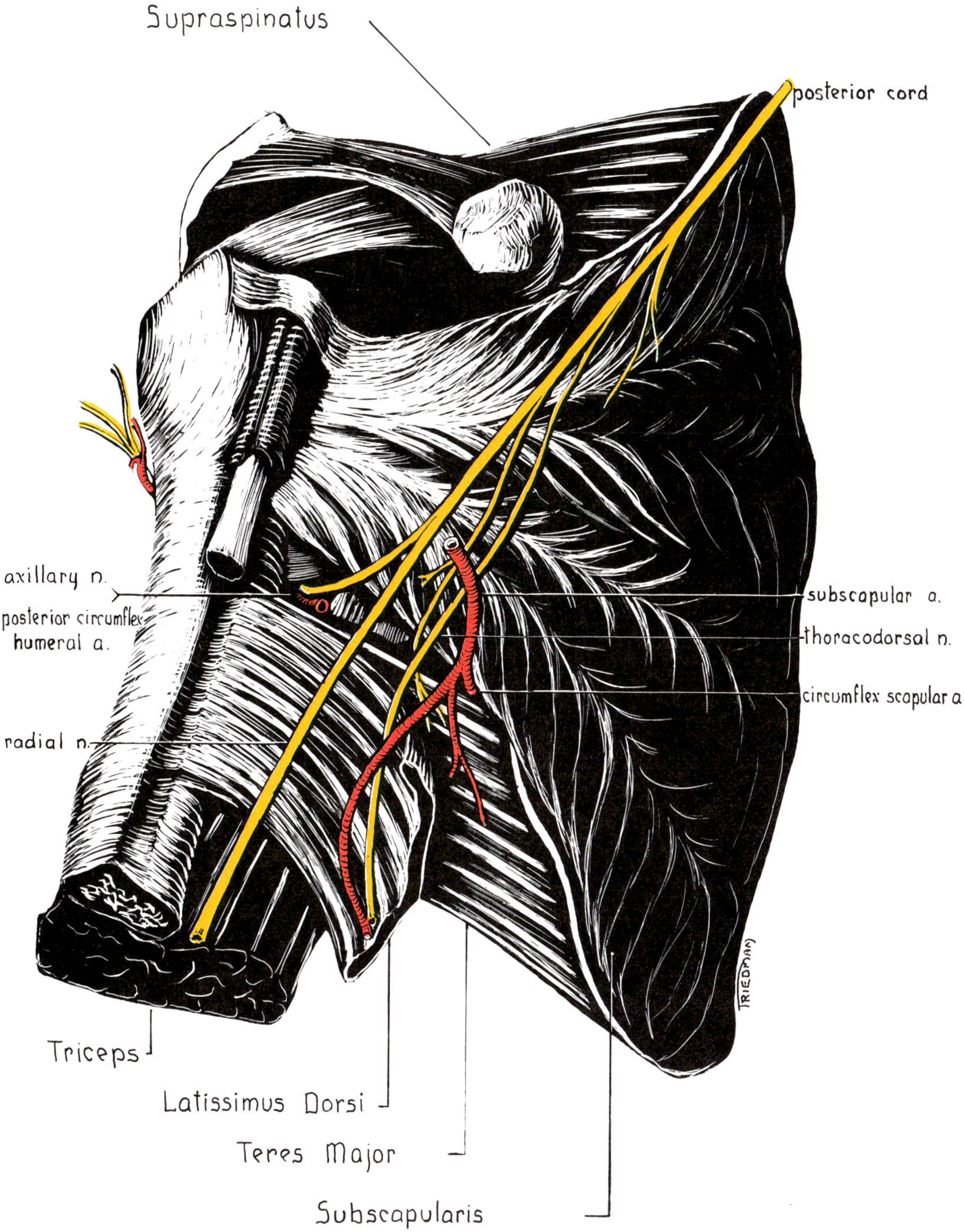

Supraspinatus
posterior cord
axillary n.
posterior circumflex humeral a.
radial n.
subscapular a.
thoracodorsal n.
circumflex scapular a.
FRIEDMAN
Triceps
Latissimus Dorsi
Teres Major
Subscapularis

Muscular Union of the Scapula and Humerus: II. Posterior Aspect

The large triangular posterior aspect of the scapula is subdivided into an upper and lower portion by the *spine,* which projects backward and then extends laterally as a cantilevered overhang, the *acromion.* This forms the *spinoglenoid notch* between the lateral margin of the overhanging spine and the back of the neck of the scapula. This aspect of the scapula is richly supplied with blood.

The *descending scapular artery* (*deep branch of transverse cervical artery*) runs downward along the medial border, sending branches into the supraspinous fossa above the spine and the infraspinous fossa below it, as well as branches into Subscapularis in front. The *suprascapular artery,* a branch of the thyrocervical trunk of the subclavian, passes over the *transverse scapular ligament,* which bridges the scapular notch, and down to supply first the supraspinous fossa and then, passing through the spinoglenoid notch, the infraspinous fossa. It is accompanied throughout its course by the *suprascapular nerve,* derived from C5 and C6 through the upper trunk of the brachial plexus; this nerve arrives on the back of the scapula by passing through the scapular notch. Lastly, as we have seen before, the *circumflex scapular artery,* a branch of the subscapular artery, reaches the back of the scapula by winding backward through the triangular space.

The relationship of the triangular and quadrangular spaces to the tendon of the long head of Triceps is clearly shown in this illustration. As already described, the quadrangular space is bounded by Triceps medially, surgical neck of the humerus laterally, Subscapularis above, and Teres Major below. It serves as a muscular interspace which permits the passage of the *axillary nerve* and the *posterior circumflex humeral artery.* A branch of the posterior circumflex humeral artery passes downward along the back of the humerus to join the profunda brachii artery lower down. This branch may on occasion replace the profunda artery.

The origin of **Teres Major** from an oval, flat area on the lower third of the axillary portion of the scapula is seen here. From this origin the muscle passes laterally in front of the long head of Triceps to insert with Latissimus Dorsi on the medial lip and floor of the intertubercular groove. Observe that the long head of Triceps joins, below Teres Major, with the lateral head arising from the upper posterolateral portion of the shaft of the humerus.

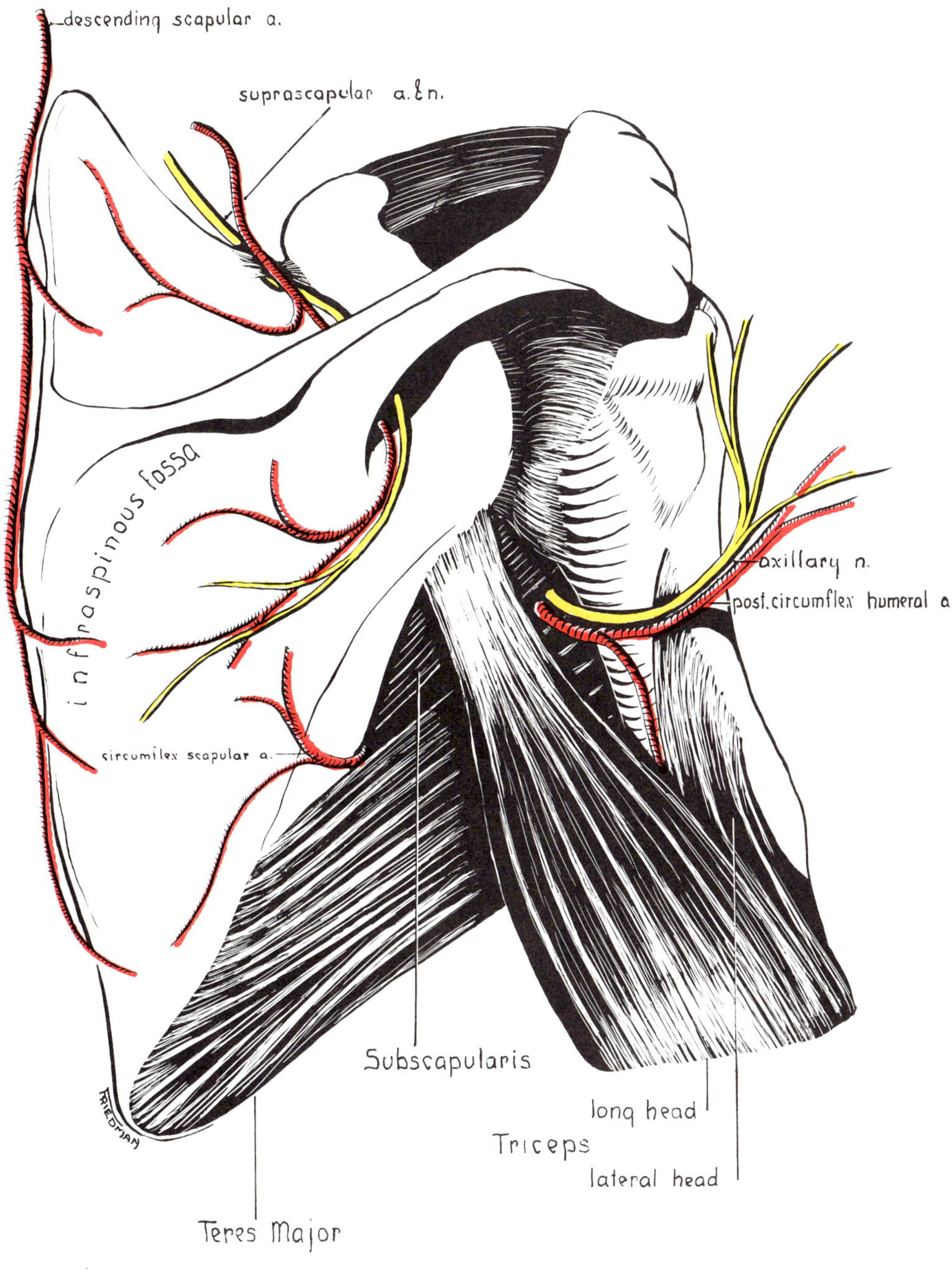
descending scapular a.
suprascapular a.& n.
infraspinous fossa
axillary n.
post. circumflex humeral a.
circumflex scapular a.
Subscapularis
long head
Triceps
lateral head
Teres Major
FRIEDMAN

Muscular Union of the Scapula and Humerus: III. Posterior Aspect (cont'd)

Infraspinatus arises from the infraspinous fossa. Like Supraspinatus above, it is thick and fleshy at first and narrows rapidly to a flat tendon. This passes behind the capsule of the shoulder joint to insert onto the middle facet surmounting the greater tubercle of the humerus. Infraspinatus receives its nerve supply from the suprascapular nerve, which lies deep to it.

Teres Minor, a relatively small, inconspicuous muscle, arises from a flat area on the back of the scapula just above Teres Major. It runs laterally, parallel with and almost indistinguishable from Infraspinatus to insert just below this muscle on the lowest facet of the greater tubercle of the humerus. Teres Minor receives its nerve supply from the axillary nerve, which passes below it into the back of the arm.

The **"Rotator Cuff,"** the group of muscles encircling the glenohumeral articulation, has now been completely defined. In résumé it consists of Subscapularis in front, matched posteriorly by a pair of muscles, Infraspinatus and Teres Minor, and superiorly by a single muscle, the Supraspinatus. All of these muscles insert immediately beyond the joint capsule after partly fusing with it. Most have bursae underlying their tendons. The bursa under Subscapularis usually communicates with the joint; those underlying Infraspinatus and Teres Minor only occasionally communicate with it. This group of muscles acts to keep the head of the humerus in close apposition with the glenoid cavity throughout the full range of motion. The joint capsule is unsupported inferiorly when the arm is hanging at the side but becomes well supported by the long head of Triceps as the arm is abducted.

The muscles of the Rotator Cuff also function in movements of the upper limb. Infraspinatus and Teres Minor attach to the greater tuberosity and rotate the humerus externally, while Subscapularis attaches to the lesser tuberosity, and is an internal rotator. Supraspinatus inserts into the upper facet of the greater tuberosity and assists Deltoid in abduction of the glenohumeral joint. Clinically, ruptures of the Rotator Cuff, particularly of Supraspinatus, are associated with limitation of glenohumeral movement.

Teres Major passes laterally to the humerus, running from the back of the scapula to the front of the humeral shaft. Note that the quadrangular and triangular spaces, bounded below by Teres Major alone, are limited above by both the anterior and posterior parts of the Rotator Cuff. The posterior circumflex humeral artery and the axillary nerve thus pass backward below the Rotator Cuff around the surgical neck of the humerus and disappear, as will be seen, into the substance of the Deltoid. Fractures of the surgical neck of the humerus are relatively common and may involve the vessel or nerve or both, either at the time of injury or later by their incorporation into the callus.

Notice that so far the crest of the spine of the scapula and the overhanging acromion are free of muscular attachments. They are essentially subcutaneous throughout their course and readily palpated.

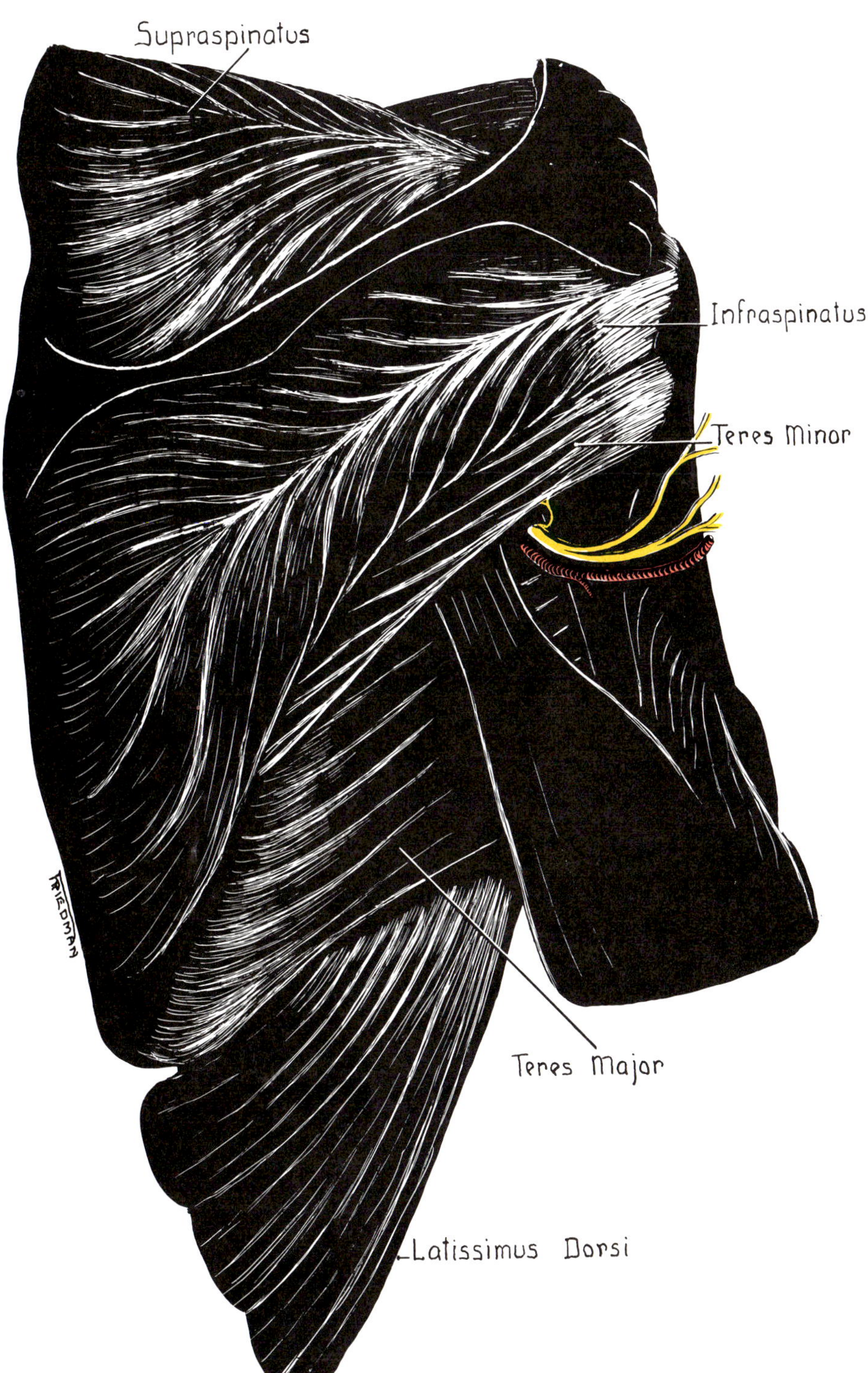
Supraspinatus
Infraspinatus
Teres Minor
Teres Major
Latissimus Dorsi
FRIEDMAN

The Suspension of the Scapulohumeral Unit: I. The Deep Group of Vertebral Muscles

The scapula is anchored to the vertebral column by an almost continuous sheet of muscle which is attached to its medial border.

Levator Scapulae arises by slips from the posterior tubercles of the transverse processes of the first four cervical vertebrae. It descends downward and backward to insert into the vertebral border of the scapula above the root of its spine. Most of its course is in the posterior triangle of the neck, whose floor it helps to form. It is innervated there by direct branches of the ventral rami of C3 and C4, which reach it through the cervical plexus.

Rhomboideus Minor arises from the spines of the 7th cervical and 1st thoracic vertebrae with extensions slightly upward to the ligamentum nuchae and downward to the supraspinous ligament. It inserts into the vertebral border of the scapula opposite the root of the spine. It is almost continuous with, and frequently cannot be separated from, **Rhomboideus Major,** which arises from the spines of the next four cervical vertebrae and the supraspinous ligaments between them and inserts into the vertebral border of the scapula between the root of the spine and the inferior angle. The Rhomboids are innervated from their deep surfaces by the *dorsal scapular nerve* (C5).

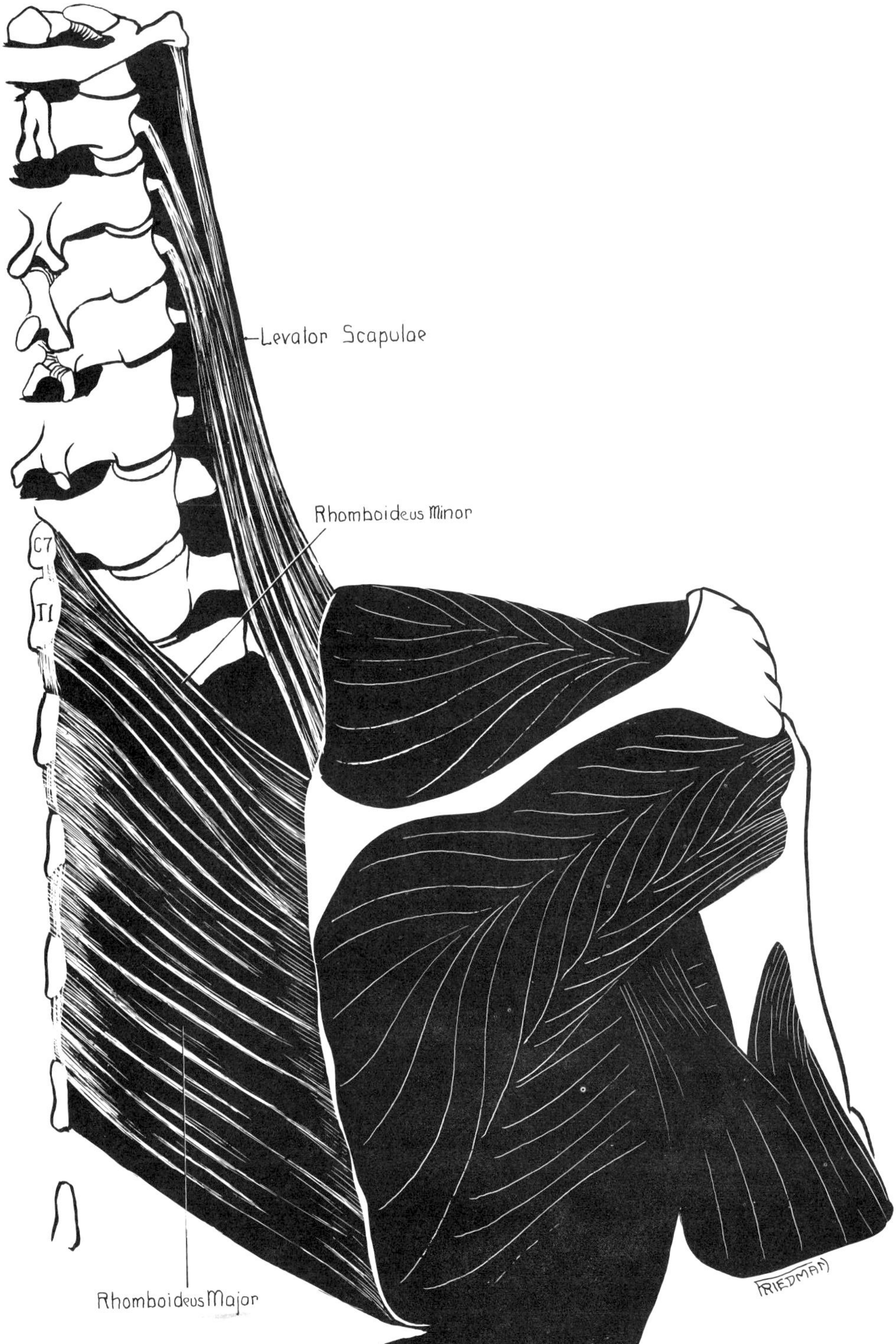
Levator Scapulae
Rhomboideus Minor
C7
T1
Rhomboideus Major
FRIEDMAN

The Suspension of the Scapulohumeral Unit: II. The Superficial Group of Vertebral Muscles

Trapezius is the prime muscle involved in the suspension of the upper limb. It arises from the medial third of the superior nuchal line, from the external occipital protuberance, from the ligamentum nuchae which anchors it to all of the cervical vertebrae, and from the spine of the 7th cervical vertebra and the following 10 or 12 thoracic spines with the supraspinous ligaments between them. From this broad origin, which is aponeurotic in its middle portion, the fibers converge into the continuous line formed by the upper border of the crest of the spine of the scapula, the inner border of the acromion, and the inner border of the lateral third of the clavicle. Trapezius thus not only suspends the scapulohumeral unit but is in a fine position to draw the arm backward and medially or to tilt the scapula upward. It is innervated primarily by the accessory nerve (XI) with supplementary proprioceptive fibers from C3 and C4.

The **Deltoid** muscle continues the pull of Trapezius. This muscle arises from the opposing aspect of the same bones into which Trapezius inserts. Thus, it attaches to the lower border of the crest of the spine of the scapula, to the outer border of the acromion, and to the anterior aspect of the lateral third of the clavicle. It is a thick, fleshy muscle which forms the prominence of the upper part of the arm. It inserts by a strong tendon into the *deltoid tuberosity* marking the lateral aspect of the middle of the shaft of the humerus. It is innervated by the axillary nerve (C5, 6).

Trapezius and Deltoid obviously are involved in the suspension of the upper extremity from the trunk. In a reverse sense, however, the upper extremity must at times support the weight of the trunk, as when one hangs from a tree branch. In this case the primary pull is transferred from the arm by way of Deltoid to the scapula and clavicle and back to the vertebral column by way of Trapezius. In addition, Latissimus Dorsi posteriorly and Pectoralis Major anteriorly are also brought into play. Both of these muscles attach to almost the same area of the humerus—Latissimus Dorsi into the floor of the intertubercular groove and Pectoralis Major into its lateral lip. Both spread the weight of the hanging body evenly from the arm to the front of the chest and to the small of the back. More commonly, both Latissimus Dorsi and Pectoralis Major are involved whenever the arm must be drawn close to the body against an opposing force. These muscles are thus the powerful movers of the arm in all exercises involving the upper limb, e.g., rowing, swimming.

Latissimus Dorsi arises from an extensive area, beginning with an aponeurotic origin from the lower thoracic spines where it is overlapped by the lower fibers of Trapezius and extending downward to involve all of the tough, dense lumbodorsal fascia (which spreads the pull down to the sacrum) and forward to the posterior part of the crest of the ilium. From here the muscle tapers rapidly to a flat tendon which passes up in front of Teres Major to fuse with it close to its insertion. Its uppermost fibers pass almost horizontally across the back and usually help to clamp the inferior angle of the scapula to the chest wall. Latissimus Dorsi is innervated from its anterior surface through the thoracodorsal nerve (Fig. 5).

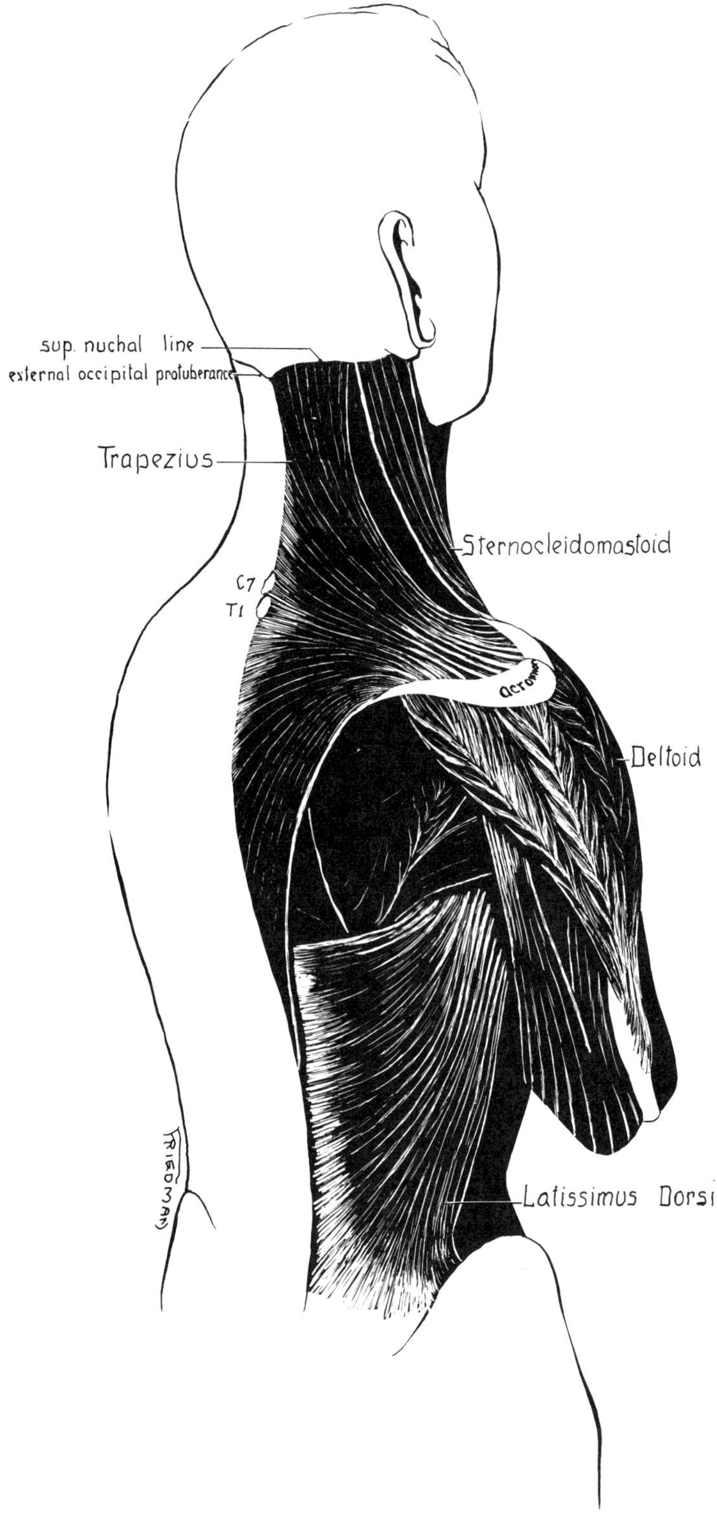
sup. nuchal line
external occipital protuberance
Trapezius
Sternocleidomastoid
C7
T1
acromion
Deltoid
Latissimus Dorsi
FRIEDMAN

The Axilla:
I. The Posterior and Medial Walls

In the course of our description of the manner in which the scapula is joined to the humerus, we have called attention to the first of the walls that form a muscular limit to the axilla and its contents. This posterior wall is made up of Subscapularis above and the conjoint portions of Latissimus Dorsi and Teres Major below. The coracoid process forms a natural upper limit for the wall, but the lower limit is less definite, since it follows the sloping lower border of Latissimus Dorsi. Laterally, the wall is limited by its attachment to the humerus, i.e., to the lesser tubercle and its crest. The anterior wall of the axilla will later be shown to be limited by the insertion of Pectoralis Major into the crest of the greater tubercle, so it follows that only a small area of the humerus, the floor of the intertubercular groove, intervenes between posterior and anterior walls of the axilla. This portion of the humerus is often defined as a fourth wall, and the tendon of the long head of Biceps as the most lateral of the axillary contents.

Medially, Serratus Anterior arises from the chest wall and sweeps backward in a half circle to insert close to the medial border of the scapula (Fig. 1). As it approaches this insertion it lies directly against the medial half of Subscapularis and leaves only a potential space between them. Thus only the lateral half of the posterior wall of the axilla actually makes any direct contact with the contents of the axilla.

Serratus Anterior originates by a series of digitations from the upper eight ribs. Acting from the thoracic wall, it holds the medial border of the scapula close to the chest wall and also pulls the scapulohumeral unit forward to increase the reach, as in fencing. It is one of the very few muscles in the body to be innervated from its external surface. The *long thoracic nerve,* which is the nerve to Serratus Anterior, arises from the ventral rami of C5, 6, and 7 while these are still roots of the brachial plexus. The nerve passes down behind the plexus and over the first rib to gain the surface of the muscle. At one time the nerve was known as the respiratory nerve because, if the scapula is held fixed in position, Serratus Anterior can be used to elevate the ribs as an aid to breathing in conditions involving respiratory embarrassment. If the nerve is interrupted the scapula becomes winged. The segmental innervation of Serratus Anterior is similar to that of Deltoid; this is a useful point in localizing nerve lesions.

Surmounting the shoulder joint a large bursa lies between the coraco-acromial arch and the Deltoid muscle superiorly and the Supraspinatus tendon and greater tubercle of the humerus inferiorly. This is the **subacromial** or **subdeltoid bursa.** It is commonly the site of inflammation and other pathological processes which may extend downward to involve the Supraspinatus tendon and the joint itself. Equally important is its accessible location; it can be successfully entered with a needle passed horizontally through the Deltoid and below the acromion.

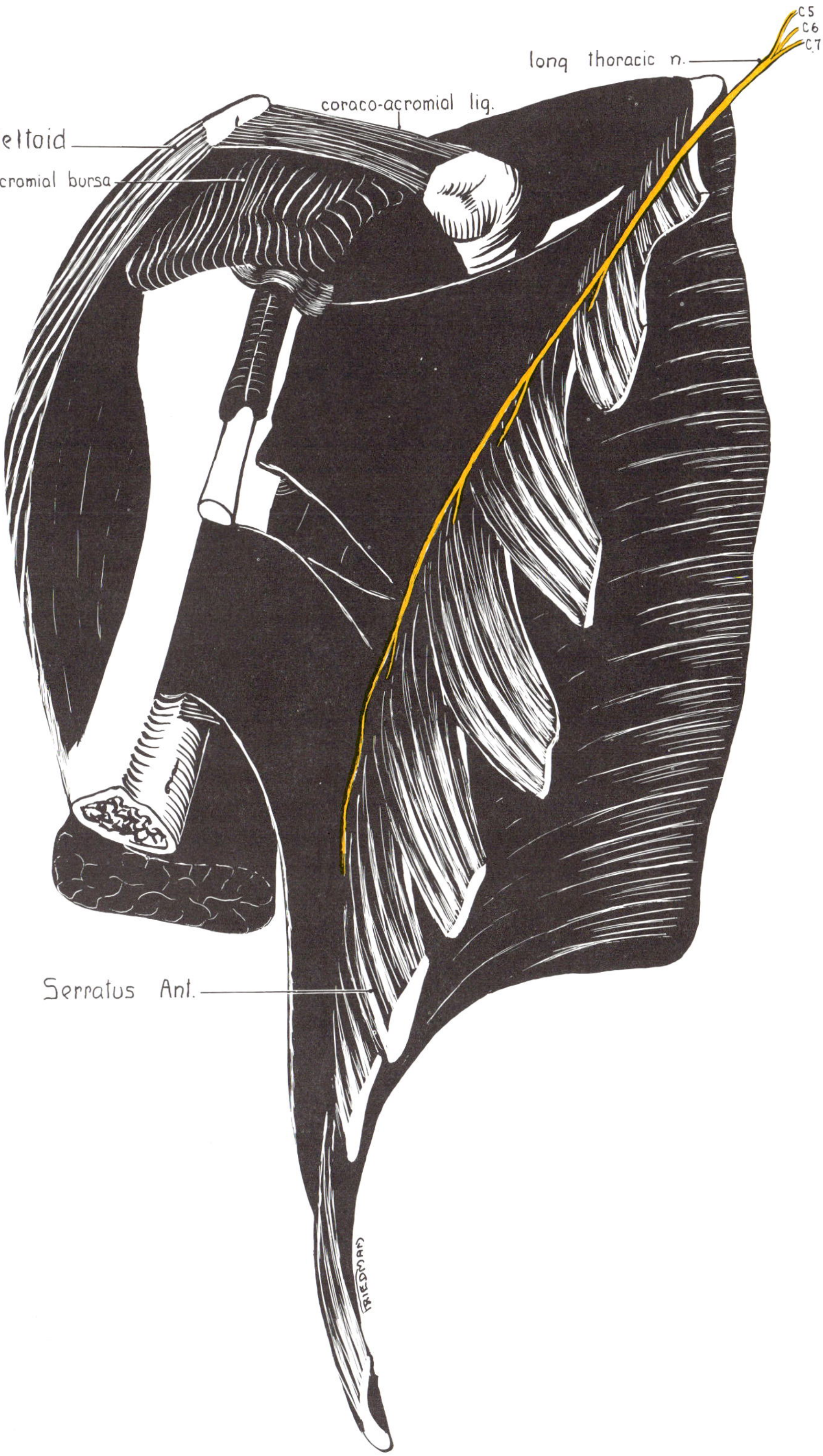
C5
C6
C7
long thoracic n.
coraco-acromial lig.
Deltoid
subacromial bursa
Serratus Ant.

The Axilla:
II. The Axillary Sheath

The major occupant of the axilla is a neurovascular bundle wrapped up in connective tissue, the axillary sheath. This **axillary sheath** is derived from the investing fascia of the neck and is carried down through the axilla until it fades out by blending with the connective tissue surrounding the individual structures as these move off to their specific tasks. The neurovascular bundle begins at the lateral border of the first rib and ends at the lower border of Latissimus Dorsi, where it leaves the axilla to enter the arm lying in front of Triceps. It arches gently as it passes diagonally from the upper medial part of the axilla to the lower lateral part.

The venous channels of the neurovascular bundle are most medially placed. They begin as a series of individual channels, as many as six or more, at the lower border of the axilla; these unite progressively to form one large **axillary vein** usually at about midaxilla. The axillary vein is renamed *subclavian vein* as it passes to the upper surface of the first rib to lie in a shallow groove in front of the tendon of insertion of Scalenus Anterior. In the neck, the subclavian vein joins with the *internal jugular vein* to form the *brachiocephalic vein*. Where multiple veins exist, the vein closest to the medial side of the artery is usually considered the axillary vein and the others as tributaries. No further attention will be paid to the individual veins that drain the structures of the axilla itself. These are large and numerous but sufficiently redundant that they can be surgically interrupted with impunity.

The **axillary artery** is the continuation of the *subclavian artery* which passes across the first rib in a shallow groove behind Scalenus Anterior (*Visual Anatomy I: Head and Neck*). It runs through the axilla as a single large component of the axillary sheath and leaves to lie on Triceps as the *brachial artery*. Although the brachial artery does not usually divide until it reaches the elbow, it may on occasion be split in two as high as the midaxilla; and then, of course, the axillary artery appears duplicated for part or all of its course.

The primary nerve channels of the neurovascular bundle begin as the *upper, middle,* and *lower trunks of the brachial plexus* just above and behind the subclavian artery. Through a series of recombinations (Fig. 14) which occur at or close to the lateral margin of the first rib, these are reshuffled to form the *lateral, medial,* and *posterior cords of the brachial plexus* as they enter the axilla.

It is worth noting that because of the position at which the neurovascular bundle enters the axilla in relation to the curve and slope of the first rib, its component structures at first lie in front of the first rib, Serratus Anterior, and the muscles of the first intercostal space. They then move laterally and downward to lie in front of the posterior wall of the axilla and lateral to Serratus Anterior.

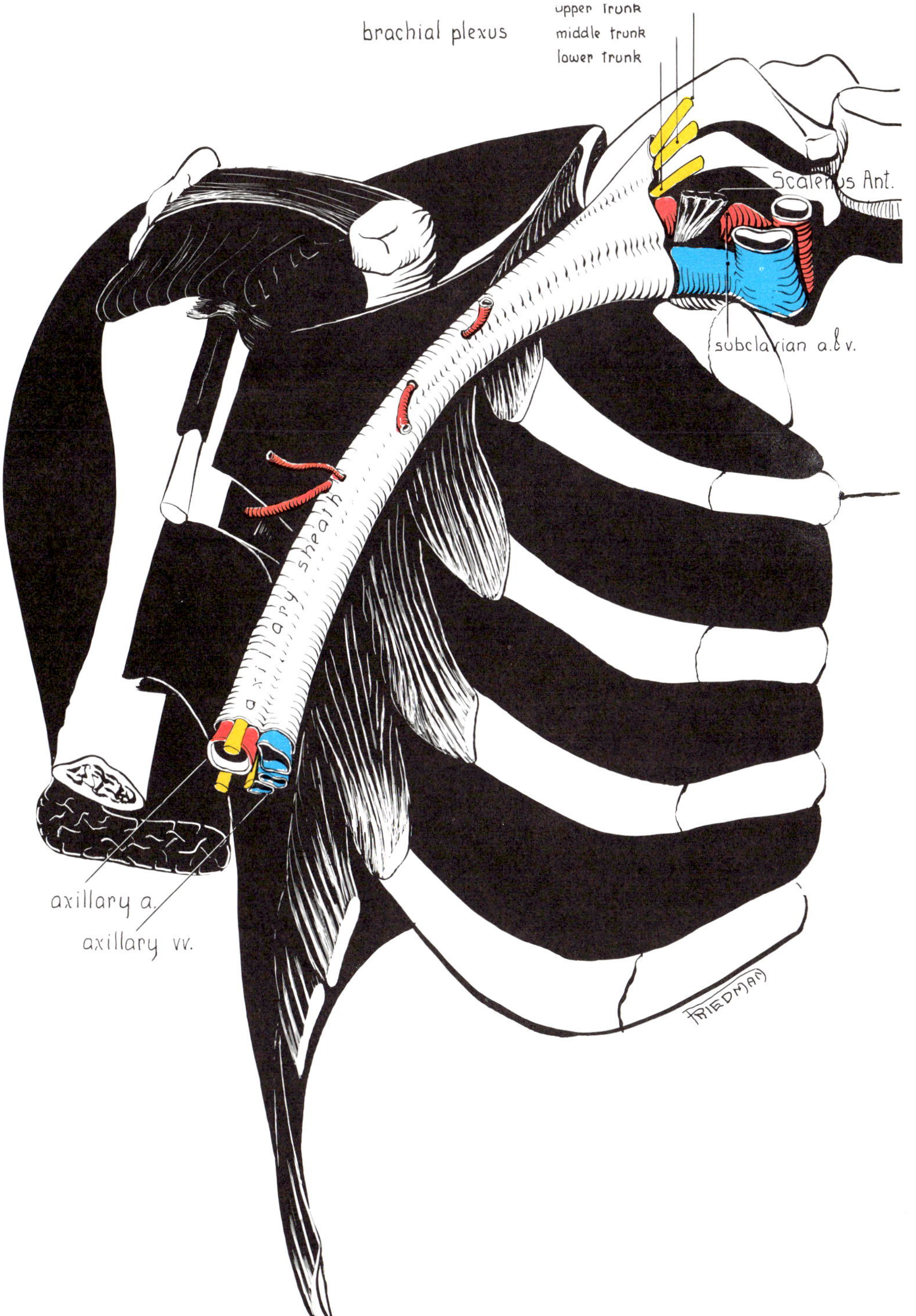
brachial plexus
upper trunk
middle trunk
lower trunk
Scalenus Ant.
subclavian a.&v.
axillary sheath
axillary a.
axillary vv.
FRIEDMAN

The Axilla:
III. The Anterior Wall

The most direct anterior limit for the axillary sheath, and indeed for all of the individual structures lying within the axilla, is a musculofascial sheet consisting of two muscles and the fascia between them. This unit extends downward from below the clavicle to the floor of the axilla, i.e., to the fascia underlying the skin of the armpit, and laterally to the flexors of the arm.

Subclavius arises by a stout tendon from the first rib close to its junction with its costal cartilage. It extends upward and laterally to insert into a groove on the undersurface of the middle of the clavicle. The fascial sheath that surrounds it continues down as a single dense membrane which finally splits to enclose Pectoralis Minor. **Pectoralis Minor** arises by fleshy slips from the 3rd, 4th, and 5th ribs and passes upward and laterally to insert as a flat tendon into the anterior aspect of the coracoid process. The membrane between Subclavius and Pectoralis Minor is termed the *clavipectoral fascia.* That portion of this fascia extending horizontally between the coracoid process laterally and the first rib medially is sometimes separately named the *costocoracoid ligament.* That part extending between Pectoralis Minor and the floor of the axilla is called the *suspensory ligament of the axilla.*

The tendon of the long head of Biceps lies between the converging posterior and anterior walls of the axilla and hence is the most lateral of its contents. The long head joins with the **short head of Biceps,** arising by dense ligament from the tip of the coracoid process, to form the definitive Biceps muscle below the lower margin of the axilla. The most medial of the muscle fibers that originate from the tip of the coracoid process become recognizably separate inferiorly as **Coracobrachialis,** which inserts on the medial side of the shaft of the humerus at about its middle, more or less opposite the insertion of the Deltoid.

The conjoint short head of Biceps and Coracobrachialis together with the tendon of the long head of Biceps form a limiting lateral wall for the neurovascular components of the axilla. By the apposition of Serratus Anterior with Subscapularis medially and by the bulky nature of Coracobrachialis and Biceps laterally, the actual free space of the axilla available to the neurovascular bundle and its branches is quite restricted.

The **clavicle** is the natural upper limit for the anterior wall of the axilla. It extends in an S-shaped curve from the acromion laterally to the sternum medially, and it articulates at both ends in freely movable diarthrodial joints. It is tightly anchored at its medial end by the *costoclavicular ligament,* which extends between the first costal cartilage and the rough undersurface of the medial end of the clavicle. Laterally it is firmly attached to the upper aspect of the coracoid process by the *coracoclavicular ligament.* This is in two parts: one, placed more anteriorly, is flat and rather square, the *trapezoid ligament;* the other, rather posteriorly placed and conical in shape, is the *conoid ligament.*

Overlying those structures of the axilla already described is the large Pectoralis Major.

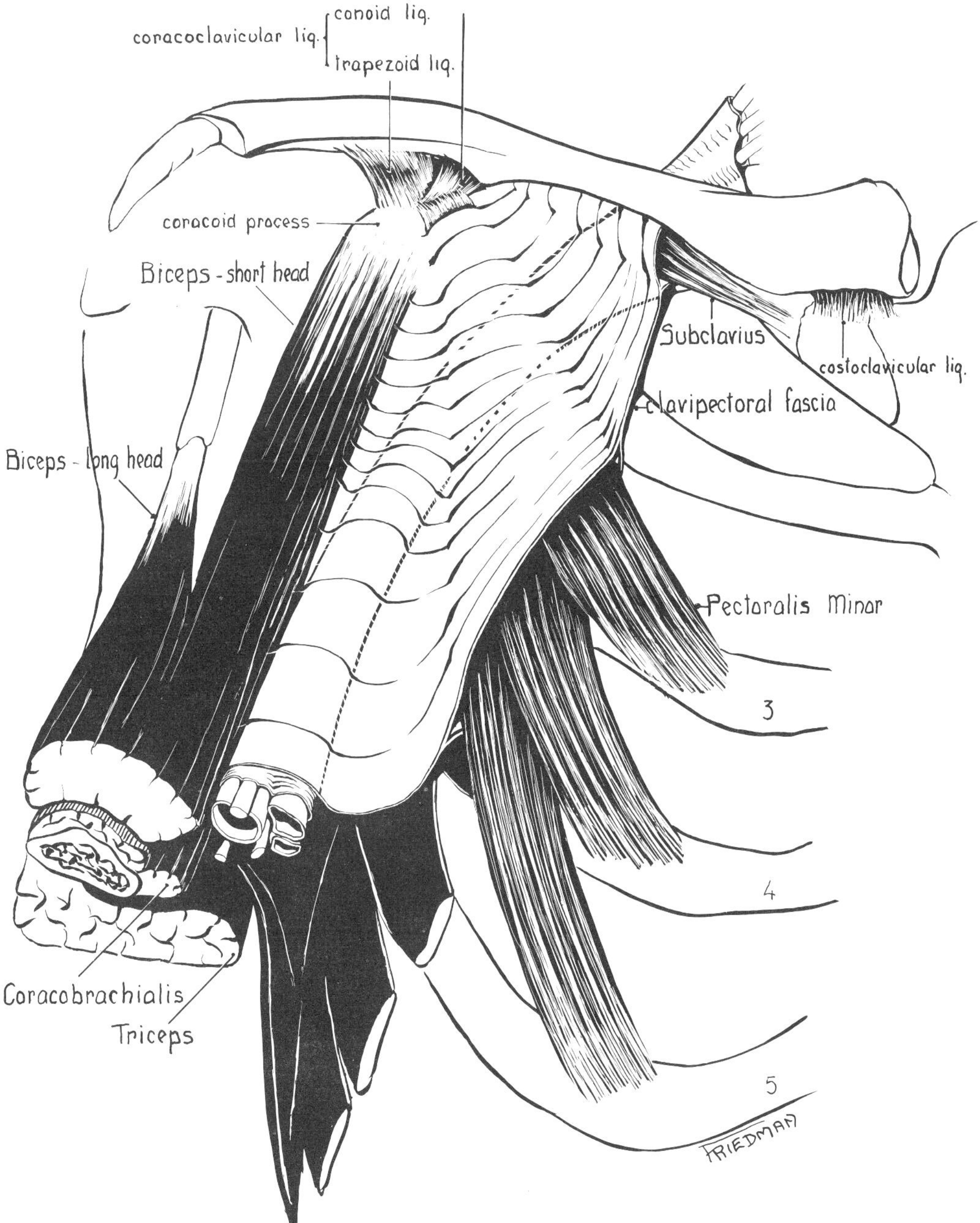
coracoclavicular lig.
conoid lig.
trapezoid lig.
coracoid process
Biceps - short head
Subclavius
costoclavicular lig.
clavipectoral fascia
Biceps - long head
Pectoralis Minor
3
4
Coracobrachialis
Triceps
5
FRIEDMAN

The Axilla:
IV. The Axillary Artery

The **axillary artery** begins at the lateral border of the first rib as the continuation of the subclavian artery (Fig. 11). It runs downward through the axilla, gradually moving lateralward to enter the arm as the brachial artery at the lower border of the conjoint Latissimus Dorsi–Teres Major tendon. It starts out lying on the muscles of the first thoracic interspace, then moves laterally on to Subscapularis and finally, as it reaches Coracobrachialis and the short head of Biceps, lies on Latissimus Dorsi and Teres Major. Anterior to the artery from above down are Subclavius, the clavipectoral fascia, and Pectoralis Minor. More superficially it is covered almost throughout its length by Pectoralis Major. It is readily palpated in the lower third of its course.

Pectoralis Minor divides the axillary artery into three parts. The first part above the muscle has only one rather inconstant branch, the *superior thoracic artery* which supplies the adjacent muscles. Two major arteries arise from the second part of the axillary artery behind Pectoralis Minor. The first, the thoraco-acromial artery, arises just under cover of the upper border of the muscle; the second, the lateral thoracic artery, just under cover of the lower border of the muscle.

The **thoraco-acromial artery** has a very short course, scarcely long enough for the vessel to pierce the overlying clavipectoral fascia before breaking up. Its branches are radially distributed in four major groups—*acromial, clavicular, pectoral,* and *deltoid*. The pectoral branches are usually the largest and most numerous since they must supply the massive Pectoralis Major muscle in addition to Pectoralis Minor. Of the deltoid branches, one runs constantly with the cephalic vein in the groove between Deltoid and Pectoralis Major on the front of the chest.

The **lateral thoracic artery** winds around the lower border of Pectoralis Minor and then follows this border to be distributed not only to the muscle but to the adjacent parts of the chest wall as well. This branch, one of the major sources of blood for the mammary gland, becomes greatly enlarged during pregnancy.

Three branches arise from the third part of the axillary artery. The **subscapular artery** (Fig. 5), is the major blood supply to the posterior wall of the axilla. It ends, along with the thoracodorsal nerve, in the substance of Latissimus Dorsi. It is a large vessel with a rather inconstant branching pattern as it distributes itself over the posterior wall of the axilla and by way of the *circumflex scapular artery* to the muscles of the back of the scapula. The other two vessels form an anastomotic ring around the surgical neck of the humerus. The larger, the **posterior circumflex humeral artery,** has been seen leaving the axilla through the quadrangular space in company with the axillary nerve (Figs. 5 and 6). As mentioned, it gives off a branch that runs down the back of the shaft of the humerus to join with the profunda brachii artery, at times replacing it. The other part of the ring, less constant and considerably smaller, is the **anterior circumflex humeral artery.** This vessel runs laterally around the front of the surgical neck of the humerus.

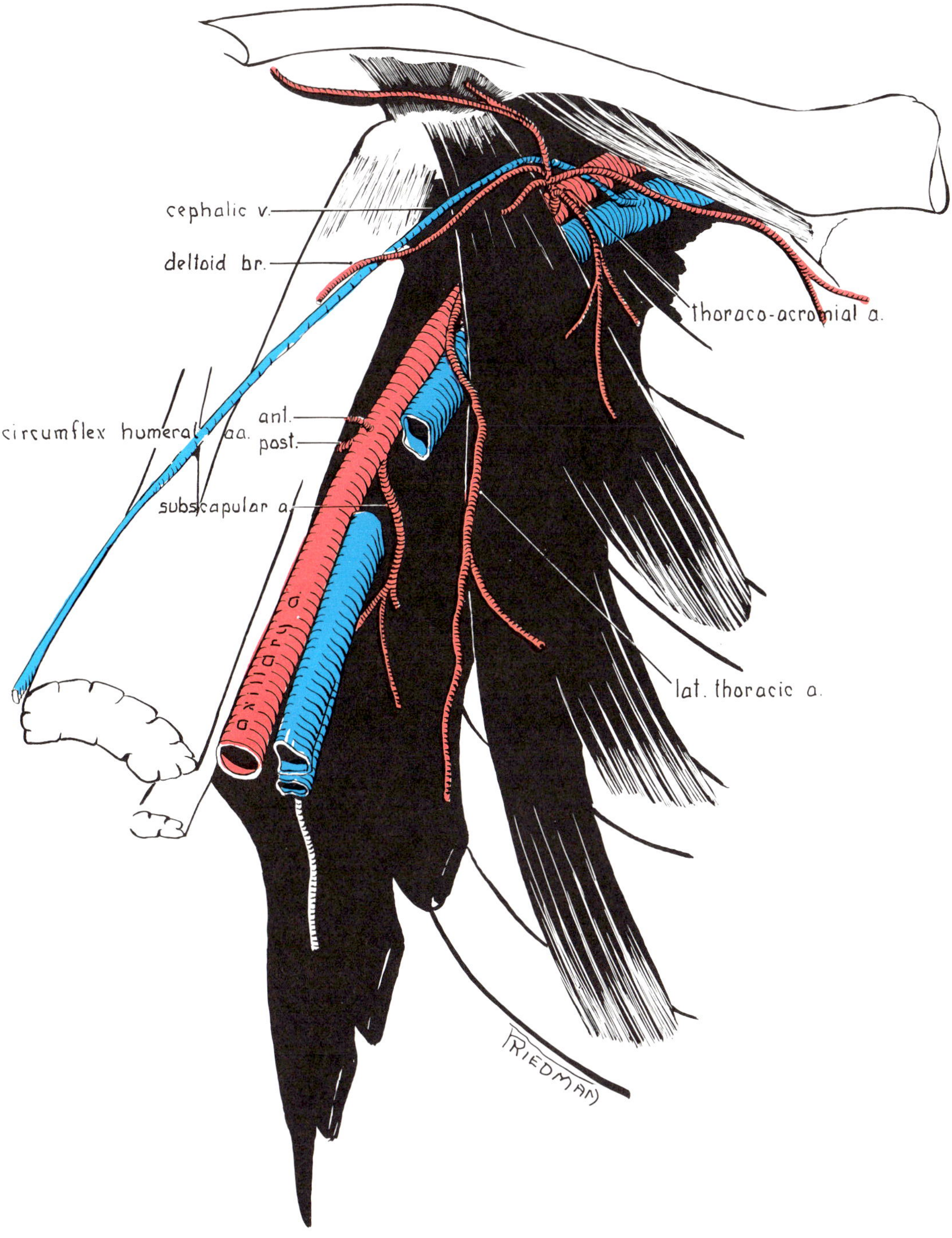
cephalic v.
deltoid br.
thoraco-acromial a.
ant.
circumflex humeral aa.
post.
subscapular a.
axillary a.
lat. thoracic a.
FRIEDMAN

The Axilla:
V. The Brachial Plexus

The **brachial plexus** begins in the neck with the ventral rami of the last four cervical and the first thoracic nerves. It also constantly receives a communication from C4 above as well as from T2 below. Arrangements in which the whole plexus is one segment higher or lower frequently occur.

The commonly described pattern of combination is as follows: the ventral ramus of C5 joins with that of C6 to form the *upper trunk* of the plexus; the ventral ramus of C7 runs alone as the *middle trunk;* the ventral rami of C8 and T1 join to form the *lower trunk.* As the plexus approaches the first rib to lie above it in front of Scalenus Medius, the fibers are reshuffled again. Some separate from each trunk as posterior divisions, and all of these recombine to form the *posterior cord* of the plexus. The remaining fibers, now termed anterior divisions, form the *lateral* and *medial cords.* Usually, the anterior division of the upper and middle trunks unite to form the lateral cord while the anterior division of the lower trunk runs alone as the medial cord.

Although there is great variation in the details of brachial plexus formation, it is as three cords that the plexus usually enters the axilla. All three start a little above and behind the first part of the axillary artery. By the time the second part of the artery is reached, the three cords are disposed around the artery according to their named positions: medial, posterior, and lateral.

Many of the branches of the brachial plexus destined for the muscles of the shoulder region arise in the neck. We have already considered the *long thoracic nerve (to Serratus Anterior)* and the *dorsal scapular nerve (to the Rhomboids),* which usually arise directly from the roots of the plexus (ventral rami), as well as the *suprascapular nerve* and the *nerve to Subclavius,* which originate from the upper trunk.

The first branches to arise from the cords of the plexus in the axilla are the lateral and medial pectoral nerves, named for the lateral and medial cords from which they originate. The *lateral pectoral nerve* passes directly forward as a leash of branches that pierce the clavipectoral fascia and enter the deep surface of Pectoralis Major. The *medial pectoral nerve,* usually single and smaller, leaves its cord a little farther down and runs forward to pierce Pectoralis Minor and then to enter the deep surface of Pectoralis Major. It supplies both muscles.

The final or terminal branches of the cords arise as they are still behind Pectoralis Minor or have passed only a short distance beyond it. The lateral cord gives off a stout *musculocutaneous nerve,* which runs laterally and downward through the substance of Coracobrachialis and deep to Biceps. The remainder of the lateral cord continues as the lateral head of the median and joins with a medial head contributed by the medial cord to form the *median nerve.* The medial cord gives off medial cutaneous nerves to the arm and forearm, then provides the medial head of the median, and terminates as the large *ulnar nerve.*

We have already seen and described the posterior cord of the plexus and its branches (Fig. 5). It gives off the *axillary nerve* and terminates as the *radial nerve.*

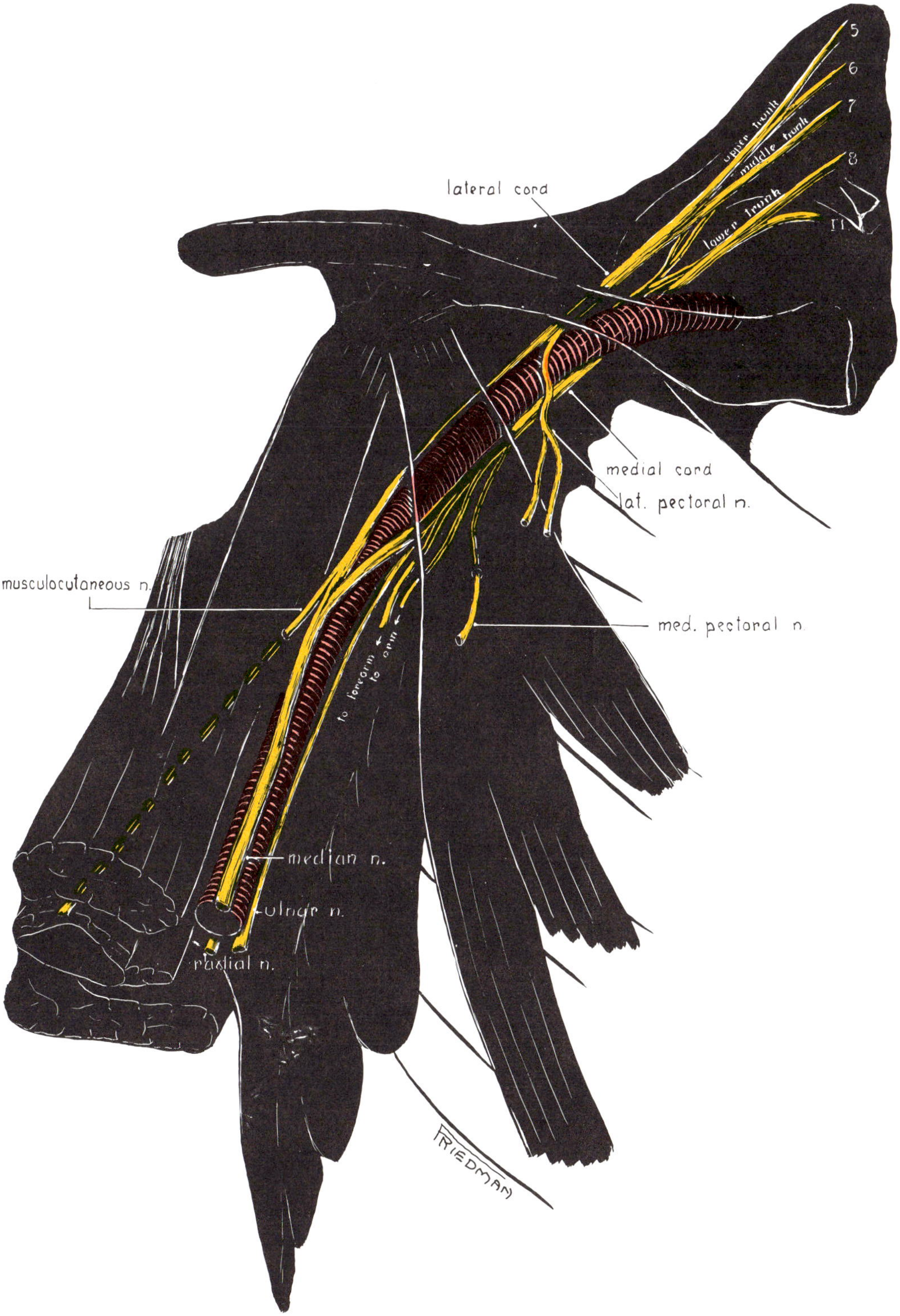
5
6
7
8
T1
upper trunk
middle trunk
lower trunk
lateral cord
medial cord
lat. pectoral n.
med. pectoral n.
musculocutaneous n.
to forearm
to arm
median n.
ulnar n.
radial n.
FRIEDMAN

The Axilla:
VI. Pectoralis Major

Pectoralis Major completes the anterior wall of the axilla. This massive muscle arises from the anterior aspect of the medial third of the clavicle, from the first six costal cartilages, from the anterior aspect of the sternum, and from the aponeurosis of External Oblique. From this extensive origin, it converges to a trilaminar flat tendon which inserts into the lateral lip of the intertubercular groove, i.e., into the crest of the greater tubercle. The muscle is innervated mainly by the *lateral pectoral nerves,* which reach its deep surface after piercing the clavipectoral fascia, and to a lesser extent by the *medial pectoral nerve,* which reaches it after piercing Pectoralis Minor. Its tendon of insertion is covered by the Deltoid muscle which caps the shoulder.

A well-defined *deltopectoral groove* contains the cephalic vein on its way to the subclavian and the deltoid branch of the thoraco-acromial artery (Fig. 13). Since the lower border of Pectoralis Major, which defines the lower border of the anterior wall of the axilla, does not extend down as far as Latissimus Dorsi, which defines the posterior wall, it follows that the lowest part of the neurovascular bundle is superficial and unprotected.

A small *deltopectoral triangle* is defined below the clavicle and between the attachments of Deltoid and Pectoralis Major. Directly underlying this is the clavipectoral fascia and the structures that pierce it. The size of the triangle varies, of course, with the muscularity of the subject. An important group of lymph nodes is located here, and on occasion these become palpable. A fuller description of these is given in Figure 44.

It will be evident by now that certain shoulder muscles readily accessible to clinical examination derive their innervation directly or indirectly from about the C5, 6, 7 level. If the musculocutaneous nerve itself is damaged, flexion of the elbow is impaired (Biceps, Brachialis); if the axillary nerve is damaged, abduction of the arm is affected (Deltoid). If the long thoracic nerve is damaged, the scapula is winged (Serratus Anterior). A lesion of the cord or of the ventral rami at the C5, 6 level, however, could easily impair the function of all three groups of muscles. A lesion a short distance distal to the ventral rami could allow Serratus Anterior to escape while impairing the function of Biceps and Deltoid.

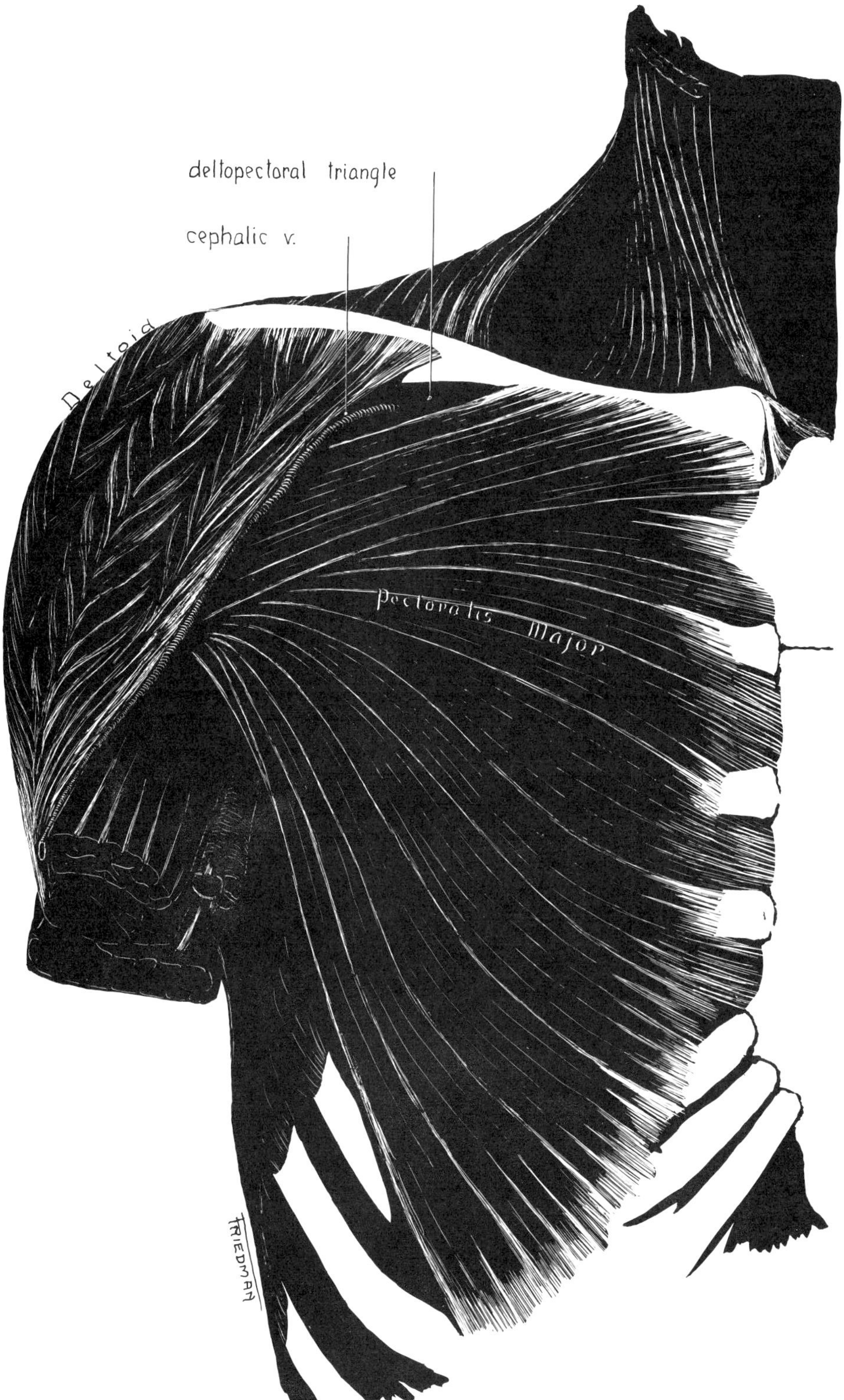
deltopectoral triangle
cephalic v.
Deltoid
Pectoralis Major
FRIEDMAN

The Arm and Elbow

The Framework:
I. The Anterior Aspect of Humerus and Ulna

The shaft of the **humerus** is roughly triangular in shape in its middle third and becomes progressively flatter as it expands toward the lower end. The anterior border, readily definable above, thus becomes less so inferiorly as the lateral and medial borders become progressively sharper to form the *lateral* and *medial supracondylar ridges* respectively. The expanded lower end of the humerus, termed the *condyle,* has knobby lateral and medial protuberances termed *epicondyles.* The articular portion of the condyle consists of a half sphere laterally, the *capitulum,* and a spool-like pulley medially, the *trochlea.* These two articular portions are indistinctly separated.

Above the capitulum is a shallow notch which is termed the *radial fossa* since it accepts the radius during full flexion. A similar notch surmounts the trochlea and is termed *coronoid fossa* since it accepts the coronoid process of the ulna.

The shaft of the **ulna,** roughly triangular in shape, expands at its superior extremity into two processes. Anteriorly, the *coronoid process* is short and its anterior surface is roughened by the tendon of Brachialis which it receives. Posteriorly, the *olecranon* extends upward in hooklike fashion to embrace the back of the trochlea and so hook on to the humerus. The semicircular *trochlear notch* formed by the coronoid process and olecranon of the ulna articulates with the trochlea of the humerus.

The primary union between the forearm and arm involves the ulna and the humerus. The bones meet one another at an angle so that when the hand is held in the supine position the ulna diverges by about 15 degrees from the body. This is termed the "carrying angle."

The secondary union in the elbow joint involves the radius, which is not shown here. The primary articulation of this bone is with the ulna, which accepts the margin of the head of the radius on a shallow radial notch, as will be seen. To be sure, the superior surface of the head of the radius does articulate with the capitulum, but the union is not a bearing joint and resection of the head of the radius need not impair the function of the elbow.

Below the radial notch of the ulna is a rough large triangular area which provides an origin for the Supinator muscle.

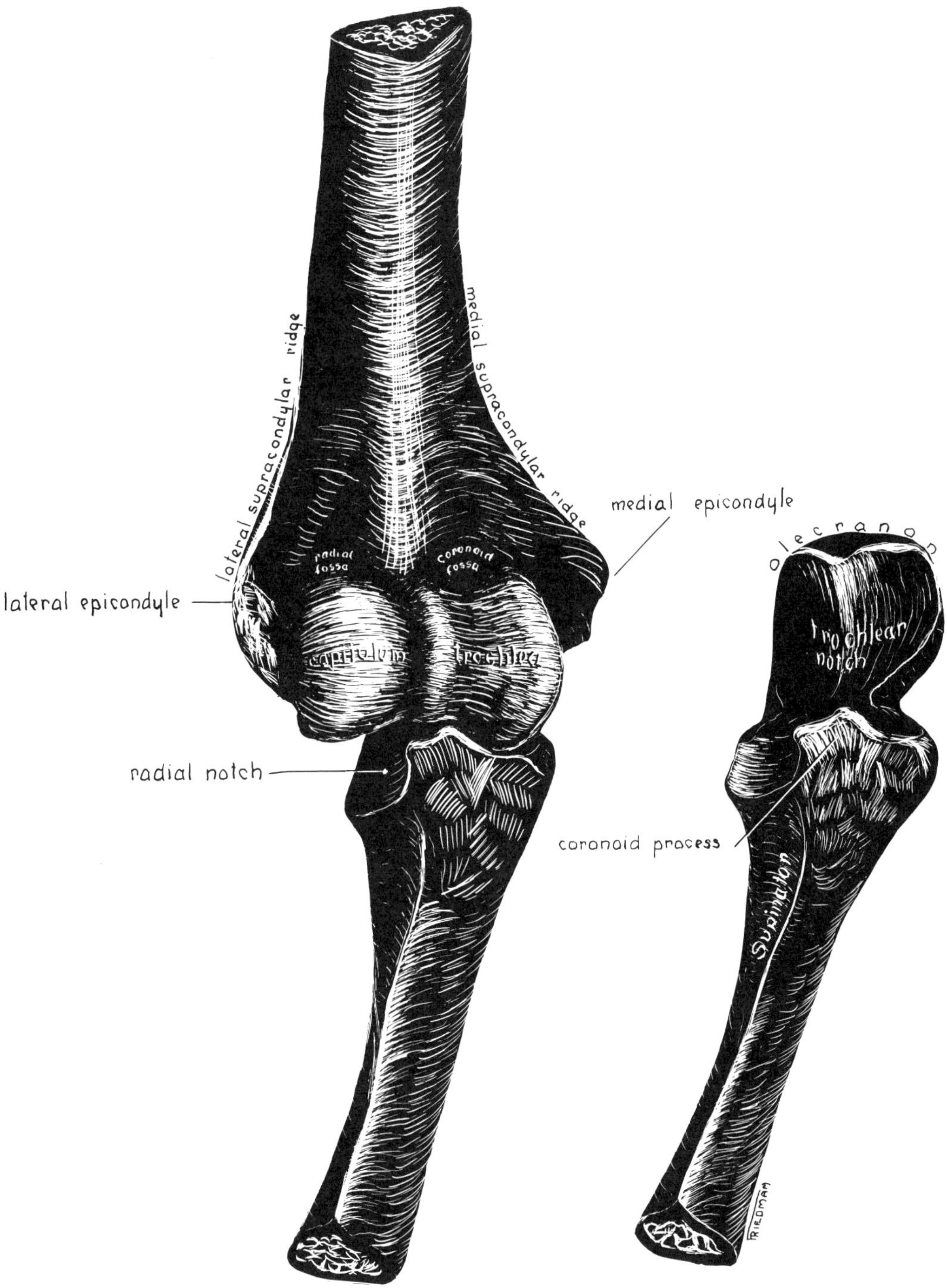
lateral supracondylar ridge
medial supracondylar ridge
medial epicondyle
radial fossa
coronoid fossa
lateral epicondyle
capitulum
trochlea
olecranon
trochlear notch
radial notch
coronoid process
supination

The Framework:
II. The Posterior Aspect of Humerus and Ulna

Casual inspection of the posterior aspect of the lower end of the humerus reveals that here the articular portion of the condyle consists only of trochlea and that no capitulum is visible. There is, of course, no need for the articular surface for the radius to extend up behind the humerus, since the needs of the radius are quite satisfied by the capitulum, which accommodates the fully extended position of the forearm. Instead, on the posterior aspect of the humerus there is a broad, flat surface on the condyle lateral to the trochlea which provides an origin for the Anconeus muscle. Above the trochlea is a deep cavity, the *olecranon fossa,* which receives the hooklike olecranon of the ulna. So deep is this fossa that only a thin plate of bone remains here, and it is not uncommon to find this plate perforated.

The posterior aspect of the ulna is marked by the presence of a large, flat triangular area at its upper end. This is a fully palpable surface separated from skin only by the *olecranon bursa.* Inflammation of this bursa is commonly known to orthopedic surgeons as "student's elbow." Above and alongside the subcutaneous triangular area, the bone is roughened by the insertion of the large tendon of Triceps which here passes into the bone. Extending down from the subcutaneous triangle is the sharp posterior border of the ulna, which can be palpated throughout its length from elbow to wrist. This border is particularly important, since it marks the separation of the flexor group of muscles lying medial to it from the extensors which lie laterally.

The Posterior Aspect of Humerus and Ulna — FIGURE 17

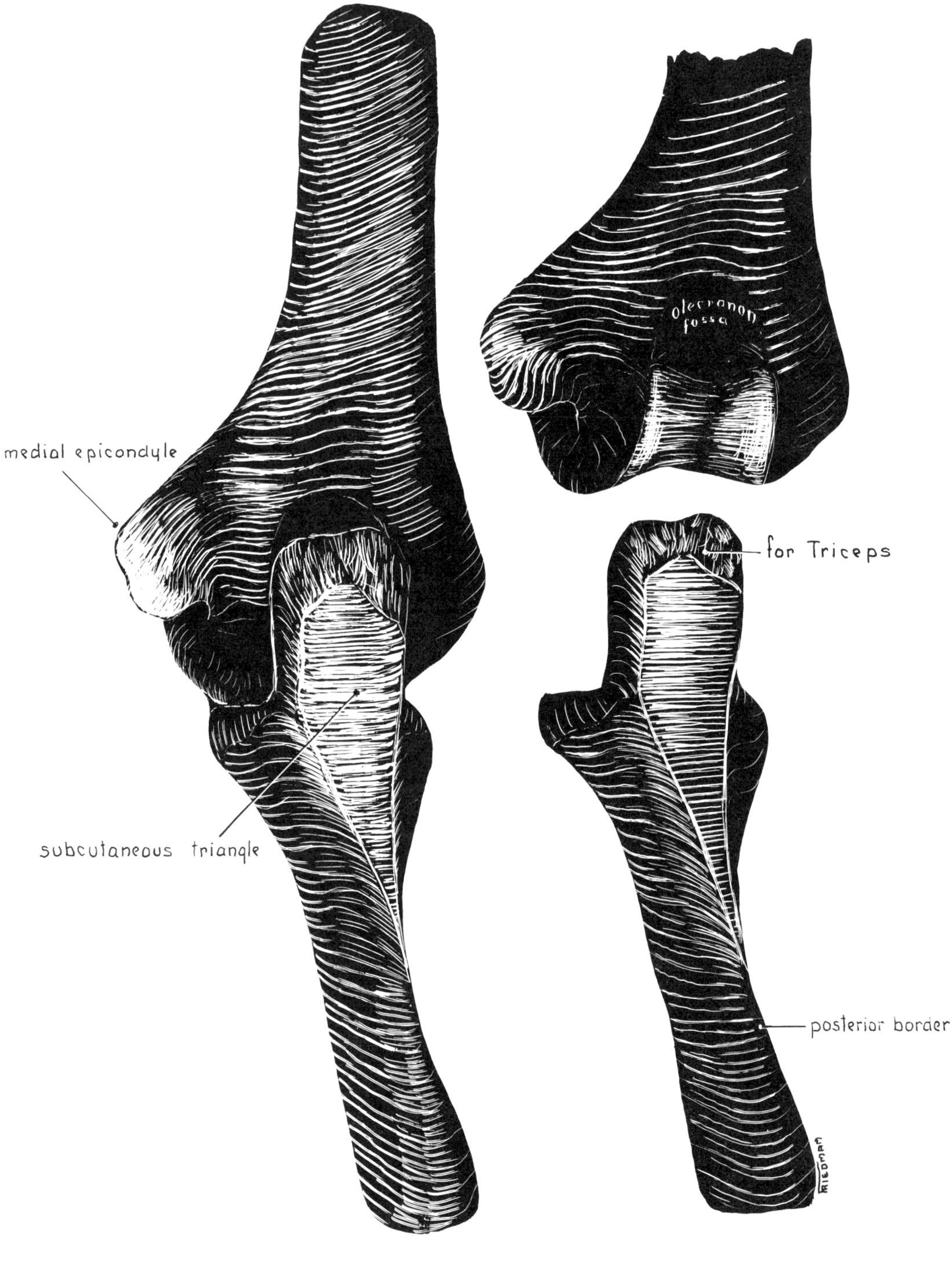

The Elbow Joint:
I. The Synovial Membrane

As illustrated in the preceding two figures, the elbow joint includes not only the primary union of humerus and ulna but the secondary union of both with the radius as well. The upper end of the **radius** expands to form a disc, the *head,* whose concave superior surface meets with the capitulum of the humerus. The capitulum presents a smooth convex surface sufficiently large to accommodate the radius through the range of movement of the elbow joint from full flexion to full extension. This requires only a half sphere on the anterior aspect of the humeral condyle. Below the head of the radius the shaft tapers for a short distance as the *neck* and then expands again beginning at the roughened tuberosity of the radius. This *tuberosity of the radius* receives the tendon of insertion of Biceps into its posterior aspect, and is separated from the tendon anteriorly by a bursa.

The primary articulation of the radius is with the ulna. The outer rim of the head is received into the *radial notch of the ulna,* and it is against this surface that the radius rotates. It is held firmly in position by a ligamentous ring, the *annular ligament,* which arises from the anterior and posterior margins of the radial notch of the ulna and encircles the head and upper portion of the neck of the radius. The annular ligament tapers in such a way as to prevent the head of the radius from being pulled down through the ring. In young children forcibly pulled out of harm's way or lifted by the hands, the radial head may subluxate into the annular ligament. This common condition is termed "pulled elbow."

For the most part, the synovial membrane follows the usual pattern. Thus, within the joint, all bones and ligaments are excluded except for the articulating surfaces, and consequently the synovial membrane is said to attach at the articular margins. Specifically, this means that seen from the front, the synovial membrane follows the margins of the articular surfaces of the condyle. It also lines both coronoid and radial fossae of the humerus before reflecting away from the bone to pass freely downward and over the trochlea and capitulum to attach to the ulna and radius. At the ulna the attachment is simple and proceeds directly to the margins of the coronoid process. At the radius, however, the membrane extends downward below the annular ligament to form a redundant pocket along the neck of the radius before returning back upward again to attach to the articular margins of the head.

Posteriorly, the attachments of the synovial membrane are unremarkable. Here, the membrane attaches to the articular margins of the trochlea; but superiorly, the synovial membrane extends upward to line the olecranon fossa first before leaving the humerus to attach to the articular margins of the ulna alongside the trochlear notch. Again we observe that the synovial membrane passes below the annular ligament and forms a pocket alongside the neck of the radius before returning to the articular margins.

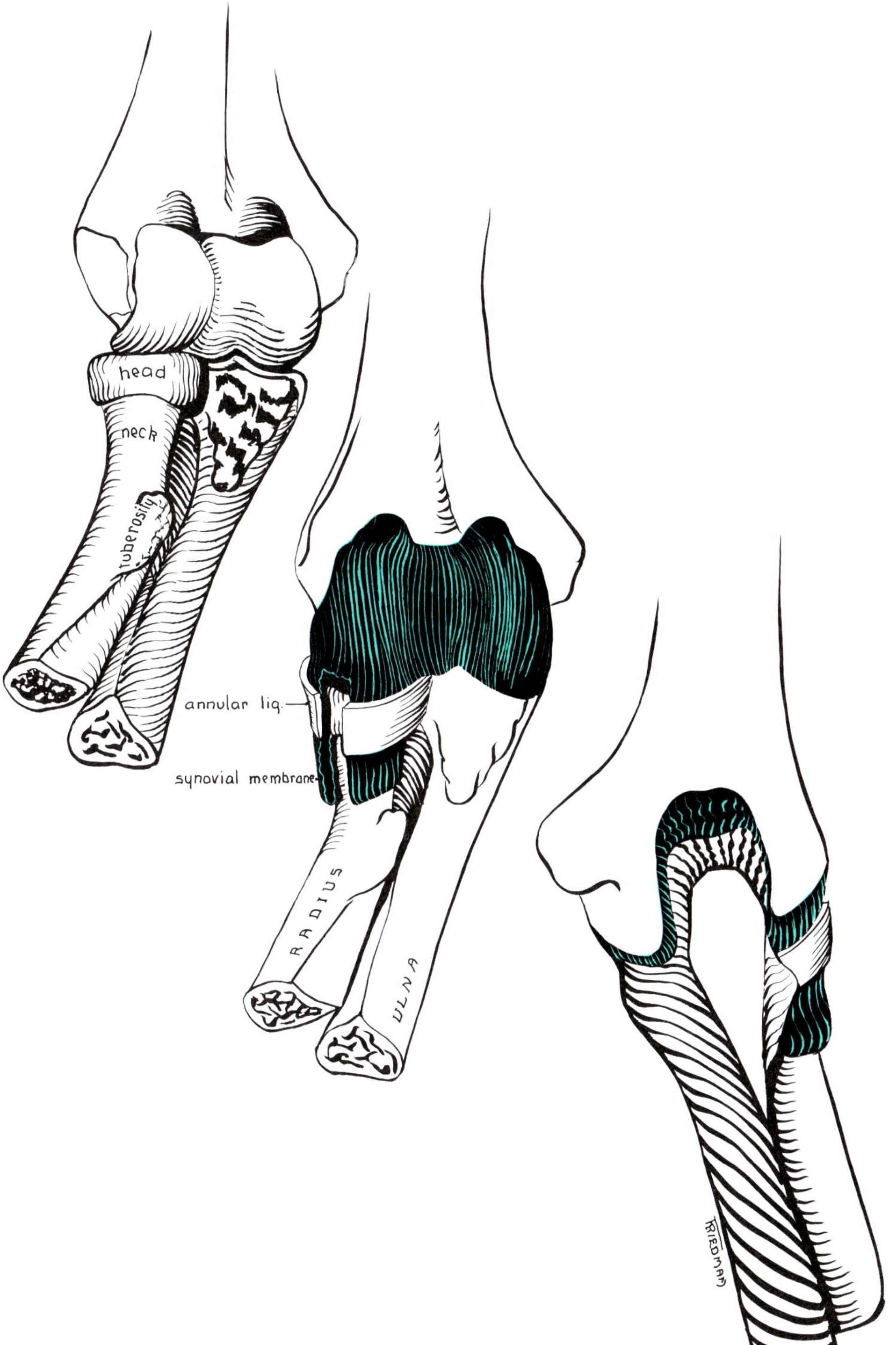
head
neck
tuberosity
annular lig.
synovial membrane
RADIUS
ULNA
FRIEDMAN

The Elbow Joint: II. The Fibrous Capsule

A simple fibrous capsular enclosure extends from the humerus above to the radius and ulna below. Superiorly, the **capsule** attaches beyond the synovial membrane just above the radial and coronoid fossae anteriorly. Behind the elbow, however, the upper margin of the fibrous capsular attachment falls short of including the whole olecranon fossa and bridges it instead. This has the effect of leaving a small portion of synovial membrane exposed; a pad of fat sits here and is sucked inward as the forearm is flexed. Inferiorly, the capsular attachment is again just beyond that of the synovial membrane on the ulna, while on the radial side it proceeds only as far as the annular ligament.

The fibrous capsule is particularly thickened laterally and medially, and specific ligaments are named. Laterally, extending between the lateral epicondyle and the annular ligament, is a ribbonlike *lateral collateral ligament*. Medially, a *medial collateral ligament* extends as a triangle between the medial epicondyle and the margins of the trochlear fossa. Three portions of this are named: an anterior band passes from the medial epicondyle to the anterior portion of the trochlear fossa, a posterior band passes from the epicondyle to the posterior portion of the trochlear fossa, and an oblique band simply provides a flange for the joint by passing from front to back of the trochlear fossa.

The lateral and medial humeral epicondyles and the supracondylar ridges above them provide attachment for the superficial extensor and flexor muscles of the forearm respectively. The area available for the attachment of these muscles is increased by fibrous intermuscular septa on either side of the humerus. These *lateral* and *medial supracondylar intermuscular septa* are shown here.

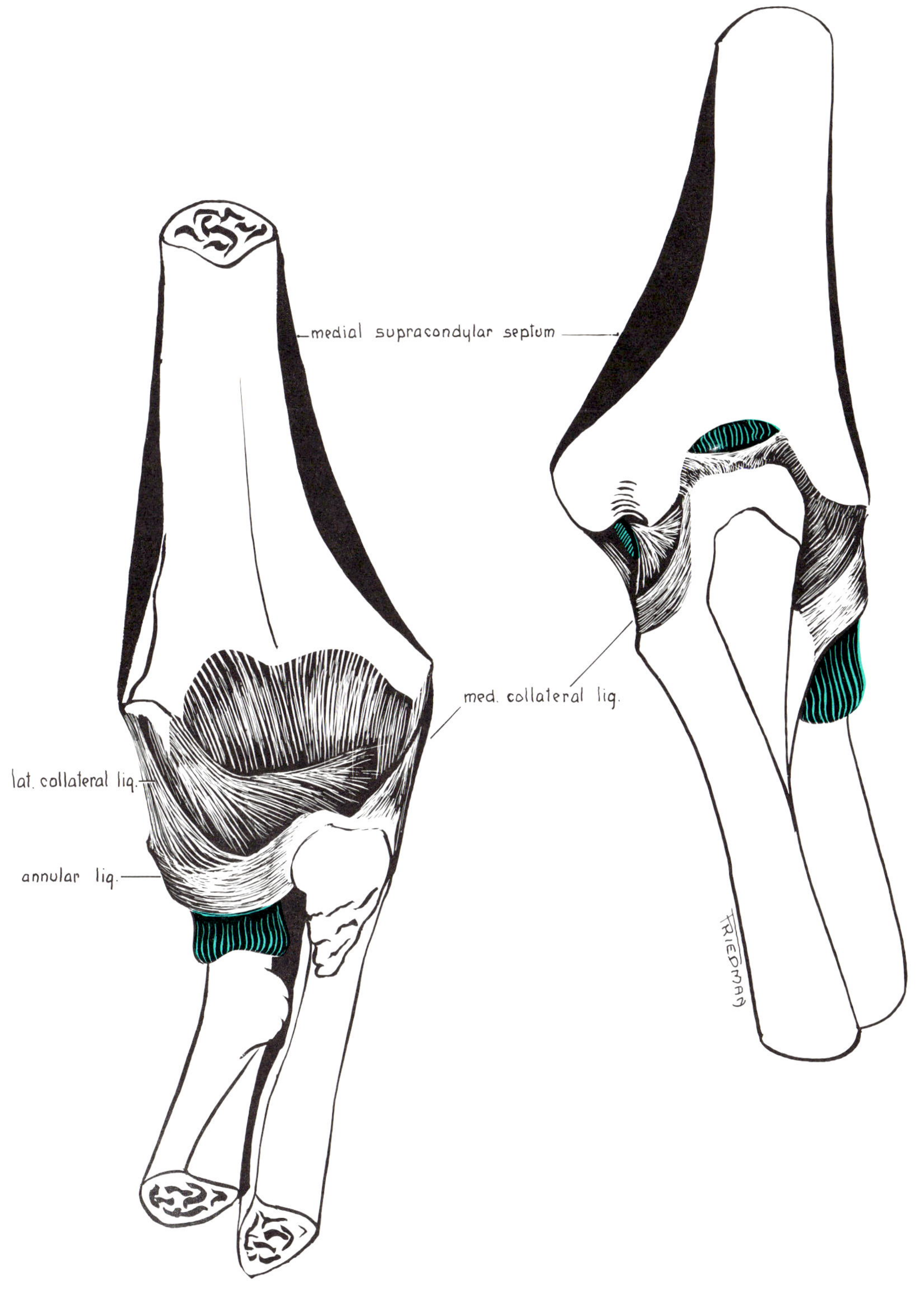
medial supracondylar septum
med. collateral lig.
lat. collateral lig.
annular lig.
FRIEDMAN

The Elbow Joint: III. Triceps and Brachialis as Stabilizers

As in the case of the shoulder joint, a cuff of muscles holds the articulating surfaces of the elbow joint in close apposition. Here, since the joint is a simple hinge which allows only flexion-extension, i.e., movement in one plane, a stabilizing cuff is required only for the anterior and posterior aspects of the joint. This function is served primarily by Brachialis anteriorly and Triceps posteriorly.

Brachialis arises from the anterior aspect of the lower half of the shaft of the humerus. Its fibers originate as high as the intertubercular groove and extend laterally and medially to the intermuscular septa. The lowest fibers originate close to the fibrous capsule of the elbow joint. Fleshy fibers give way to a flat strong tendon which passes directly in front of the capsule of the joint to insert into the anterior aspect of the coronoid process of the ulna.

Two of the heads of **Triceps** have been seen in connection with the shoulder region (Fig. 6). The *long head of Triceps* originates from the infraglenoid tubercle of the scapula and passes downward toward the shaft of the humerus where it is joined by the *lateral head of Triceps,* which originates from the back of the humerus in its upper third. Both heads pass superficial to the *medial head of Triceps,* which arises from the posterior aspect of the lower half of the shaft of the humerus and unites with the other two heads to form the tendon of insertion. The origin of the lateral head is separated from that of the medial head by a spiral, shallow *groove for the radial nerve,* which runs down and laterally across the back of the humerus. Like Brachialis in front, the medial head of Triceps extends down to within an inch of the fibrous capsule.

In Figure 14 we saw that at the lower border of the axilla the **ulnar nerve** remains in the position of its parent medial cord of the brachial plexus, i.e., medial to the axillary artery. As it passes into the arm, the ulnar nerve lies directly on Triceps, still remaining alongside the artery (now renamed brachial) until it reaches the medial intermuscular septum. Here it separates from the artery to pass behind the medial epicondyle in a groove hollowed out to receive it. It continues down behind the joint to the medial side of the ulna and gradually comes forward to the anterior aspect of the forearm.

At the lower border of the axilla we saw the **radial nerve** in the position of its parent posterior cord of the brachial plexus, behind the axillary artery. As it passes into the arm it is thus first between Triceps and the brachial artery. It quickly leaves the artery to pass downward and laterally in the shallow groove behind the middle third of the shaft of the humerus but returns again to the front of the arm at the side of Brachialis. Below the elbow joint it will again forsake the front of the forearm for the back.

In a sense, then, the radial and ulnar nerves pursue opposite courses. The ulnar nerve, trapped, as it were, behind the medial epicondyle, returns to complete its distribution from the front of the forearm and hand. The radial nerve, caught, as it were, in front of the lateral epicondyle, quickly returns to its duties in the posterior part of the forearm.

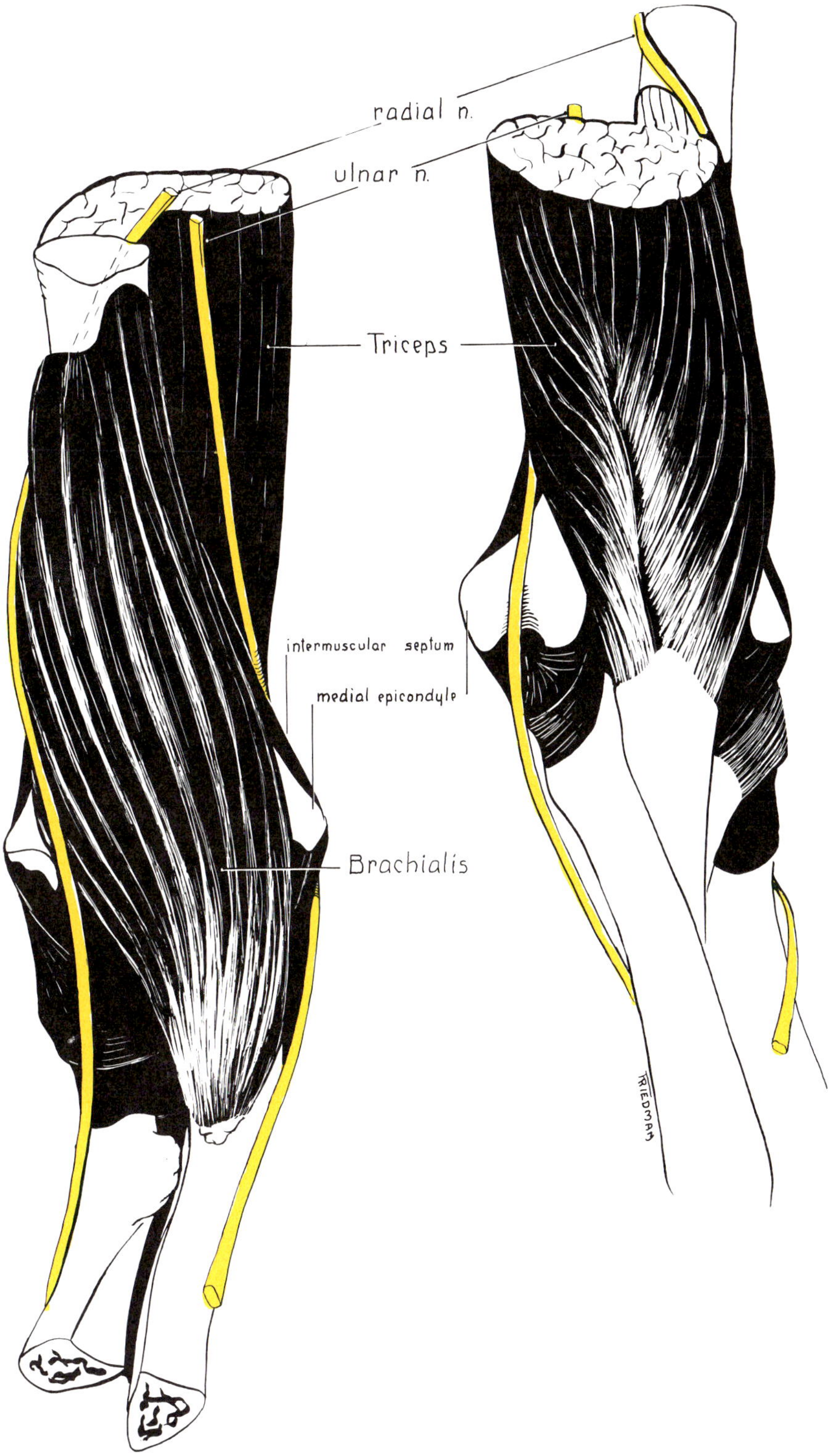
radial n.
ulnar n.
Triceps
intermuscular septum
medial epicondyle
Brachialis
FRIEDMAN

The Elbow Joint:
IV. Flexors and Extensors

Movement at the elbow is in one plane only. But even so, hard work requires powerful flexors and extensors. We have already described the Triceps muscle. Its deepest fibers, those of the medial head, are primarily involved with the matching fibers of Brachialis in stabilizing the joint. The long and lateral heads act over longer distances and may be considered as prime movers in extension. Matching these as flexors are the long and short heads of Biceps on the front of the forearm which converge to form the single strong round tendon inserting into the posterior part of the tuberosity of the radius. **Biceps** is innervated by the *musculocutaneous nerve,* which stems from the lateral cord of the brachial plexus and contains fibers derived mainly from C5 and C6. As will be seen, the tendon is also anchored to the deep fascia of the medial side of the forearm by the *bicipital aponeurosis.*

In this illustration, the radius is in the full supine position. As the forearm is pronated, the radius rolls over the front of the ulna, winding up the tendon of insertion of Biceps as it does so. Contraction of the Biceps muscle reverses this movement and so supinates the forearm; it is most effective in this when the forearm is flexed. Biceps well deserved its old name of Supinator Longus. Right-handed people use it constantly in the many tasks of the day that involve a clockwise movement. (It is a great disadvantage to be a left-handed drinker in a world of right-handed corkscrews.)

Only one other muscle is available to aid the Biceps in the task of supination, the **Supinator** muscle itself. This muscle arises in two parts. Its more superficial portion stems from the lateral epicondyle of the humerus and passes downward over the elbow joint capsule to spiral in front of the radius and insert into the oblique line on its anterior surface just below the tuberosity. The deep portion arises from a rough triangular fossa immediately below the radial notch of the ulna and, passing transversely, winds behind the radius, wraps around its lateral aspect, and inserts anteriorly into the area above the oblique line, below the capsule of the joint.

The **radial nerve** was earlier seen to pass down in a groove on the back of the humerus and then to spiral around to the front of the elbow joint. It is now seen to be overlapped by the Biceps in this part of its course. It continues down on the capsule of the elbow joint and comes to lie on Supinator. It passes to the back of the forearm between the two parts of this muscle which it innervates by twigs arising just proximal to Supinator.

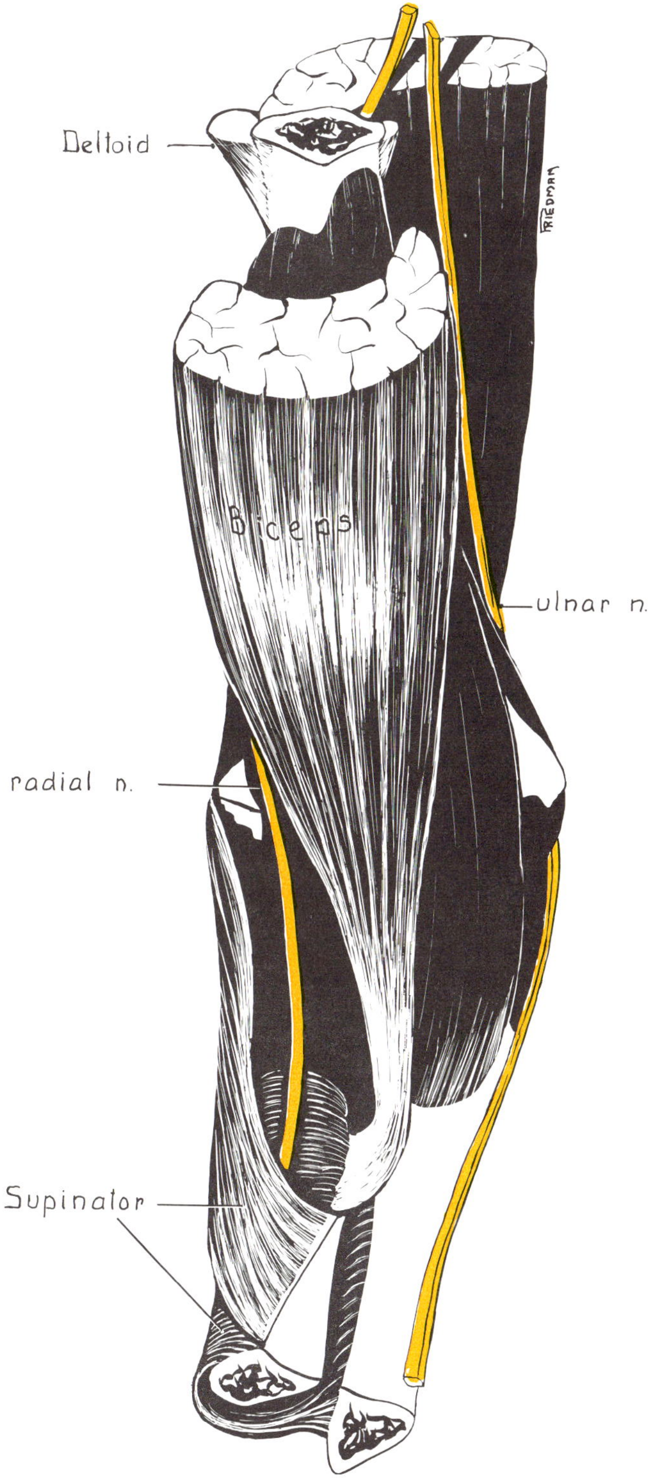
Deltoid
FRIEDMAN
Biceps
ulnar n.
radial n.
Supinator

The Elbow Joint:
V. Balance of Forces

This illustration has the sole task of emphasizing the functional balance of the muscular forces that act at the elbow joint. Flexion and extension at the elbow, and the suspension of the forearm and hand from the arm, do not involve the radius except insofar as the concave disc of the head of the radius adjusts its position with the capitulum of the humerus.

When the forearm hangs in full extension at the side of the body, Brachialis in front and Triceps behind hold the hook of the olecranon in apposition with the trochlea. In this position, since bone is hanging from bone, very little energy is expended to support the weight of the forearm. When flexion or extension are produced, the three heads of Triceps posteriorly are counterbalanced by Brachialis and Biceps anteriorly.

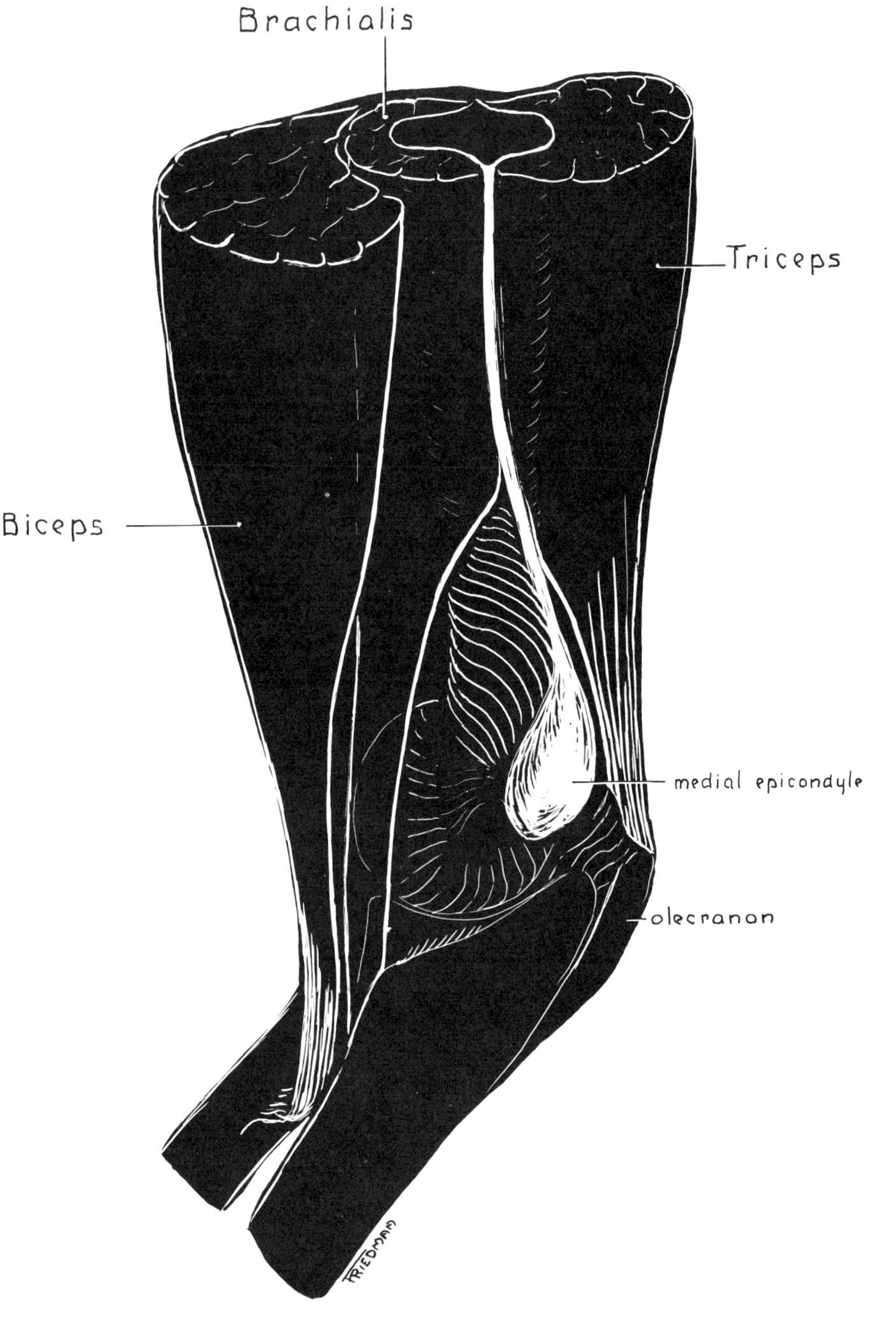
Brachialis
Triceps
Biceps
medial epicondyle
olecranon
FRIEDMAN

The Brachial Artery and Median Nerve

The **brachial artery** is the continuation of the axillary artery. It lies first on Coracobrachialis and the long head of Triceps but separates from Triceps at the medial intermuscular septum to run in front of the lower end of the humerus directly applied to Brachialis. At the level of the neck of the radius, i.e., close to the insertion of the Biceps tendon, it terminates by dividing into the *radial artery,* its more direct continuation, and the *ulnar artery.* It is at first medial to the humerus, and here one can compress it against the bone by pressing laterally. Lower down, where it lies in front of the bone, its pulse is felt by pressing backward. It is slightly overlapped by the Biceps in its upper part; but as this muscle tapers, its tendon becomes the immediate lateral neighbor of the artery at the elbow.

Of the three nerves originally disposed around the lower part of the axillary artery—the radial, ulnar, and median—only the **median nerve** accompanies the brachial artery throughout its course. This nerve begins anterolateral to the artery and gradually crosses its anterior aspect to reach the medial side at the elbow. It is useful to remember that at the front of the elbow, the Biceps tendon defines the position of the brachial artery and of the median nerve which lie, in that order, medial to it. The nerve can be blocked by an appropriate anesthetic solution introduced here.

The proximal portion of the **radial artery** descends to reach an anterolateral position on the forearm and in so doing lies first on the tendon of insertion of Biceps and then on Supinator. Before leaving Supinator it is joined on its lateral side by the rather unimportant sensory *superficial branch of the radial nerve.*

The **ulnar artery** passes somewhat medially before it too inclines more directly downward and lies on the surface of Flexor Digitorum Profundus almost throughout its course. As it reaches the medial side of the forearm it is joined on its medial side by the ulnar nerve, so that in a rather symmetrical fashion both the radial and ulnar arteries are bordered, as it were, by an accompanying nerve. The median nerve runs almost straight down the middle of the forearm. Thus it crosses the ulnar artery, from which it is separated by the few fibers of muscle that constitute the *deep head of Pronator Teres.*

There are numerous branches of the brachial, radial, and ulnar arteries in the arm and at the elbow. One class comprises the numerous branches that supply the adjacent muscles. The other class comprises branches running down from the brachial artery or up from the radial and ulnar arteries to form a rich anastomotic network about the elbow joint. Only two of these arteries warrant further detail. The *profunda brachii artery* arises from the brachial artery high up the arm and follows the radial nerve around the back of the humerus toward the lateral epicondyle to distribute itself about the front and back of the elbow joint. It also supplies branches to the muscles it passes. The *common interosseous artery* arises from the ulnar and runs downward toward the interosseous membrane which unites radius and ulna. Here it divides into *posterior* and *anterior interosseous branches* which are for the most part distributed to the adjacent muscles.

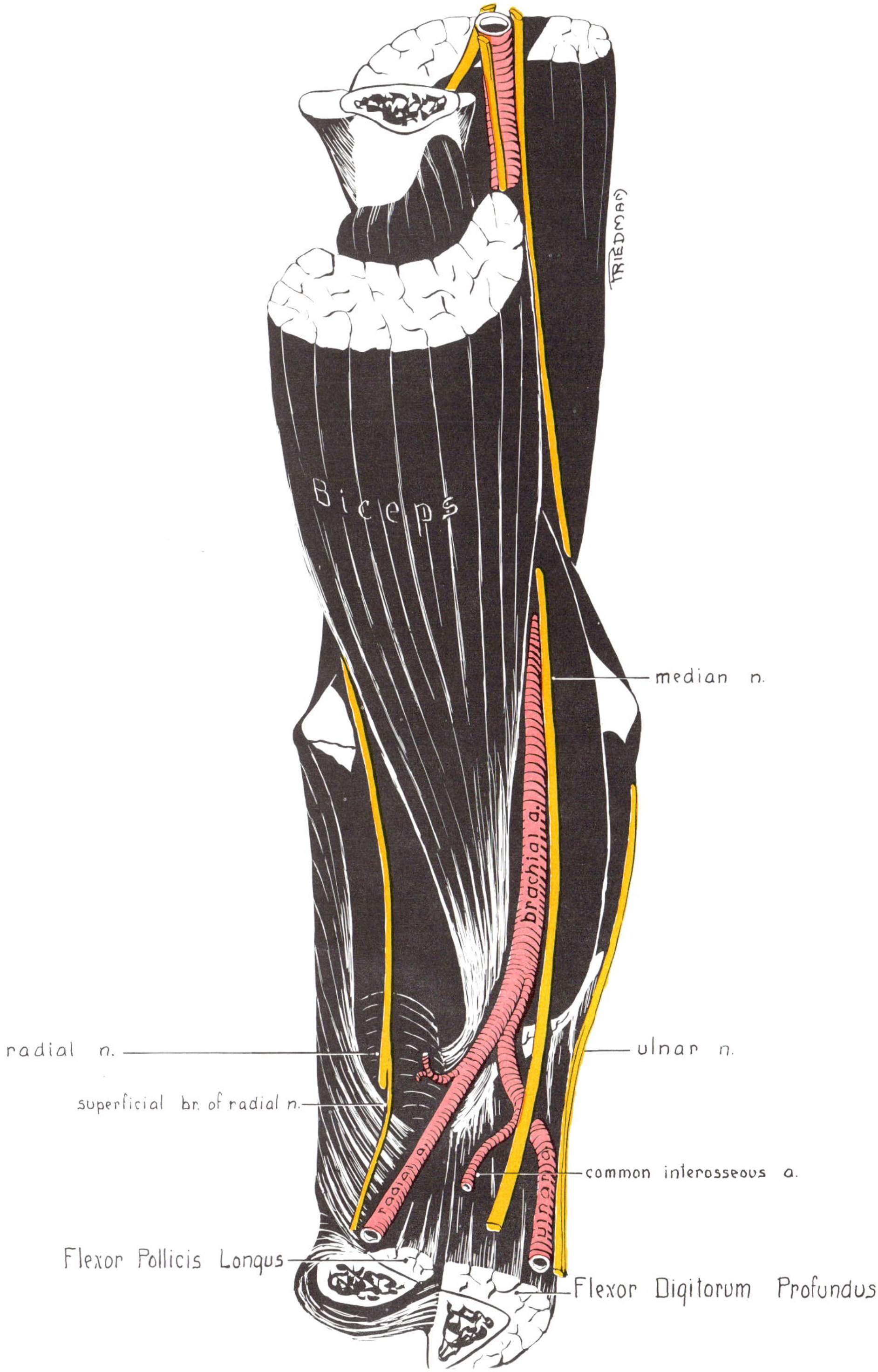
FRIEDMAN
Biceps
median n.
brachial a.
radial n.
ulnar n.
superficial br. of radial n.
radial a.
ulnar a.
common interosseous a.
Flexor Pollicis Longus
Flexor Digitorum Profundus

The Cubital Fossa

It will be appreciated that when the forearm is hanging freely at the side with the thumb pointing forward, the lateral epicondyle lies immediately above the posterior or extensor aspect of the forearm while the medial epicondyle is similarly placed with respect to the anterior or flexor aspect. A large and bulky group of superficial extensor muscles arises from the lateral epicondyle and from the supracondylar ridge immediately above it. Similarly, a bulky group of superficial flexors arises from a common flexor origin on the medial epicondyle and from the supracondylar ridge immediately above that. These groups of muscles are twisted out of their relaxed alignment when the forearm is placed in the supine or anatomical position, which is not the natural position of rest.

In the anatomical position, the flexor group from the medial epicondyle passes diagonally from its point of origin down over the front of the wrist, while the extensor group which is anteriorly placed at the elbow spirals down the lateral and posterior aspect of the forearm and hand. Dense fascial enclosures for the superficial flexor and extensor muscle groups bring them into close apposition in the upper part of the anterior aspect of the forearm. Here the most medial member of the extensor group, Brachioradialis, is contiguous with the most lateral member of the flexor group, Pronator Teres. The triangular space formed by these two muscles meeting inferiorly, with a base along a line projected between the two epicondyles, is termed the **cubital fossa.** The floor of this fossa is formed by Brachialis and Supinator and its roof by only the dense fascia of the forearm and by the *bicipital aponeurosis,* which ties the Biceps tendon to the fascia overlying the medial flexor group.

The cubital fossa contains the tendon of insertion of Biceps with first the brachial artery and next the median nerve on its medial side. In the lateral portion of the cubital fossa, covered by Brachioradialis, is the radial nerve. The termination of the brachial artery and its division into radial and ulnar arteries occur close to the apex of the triangle. The veins of the forearm coalesce to form major channels in front of the cubital fossa. They lie in the subcutaneous tissue superficial to the dense fascial roof of the cubital fossa and are readily available for venipuncture.

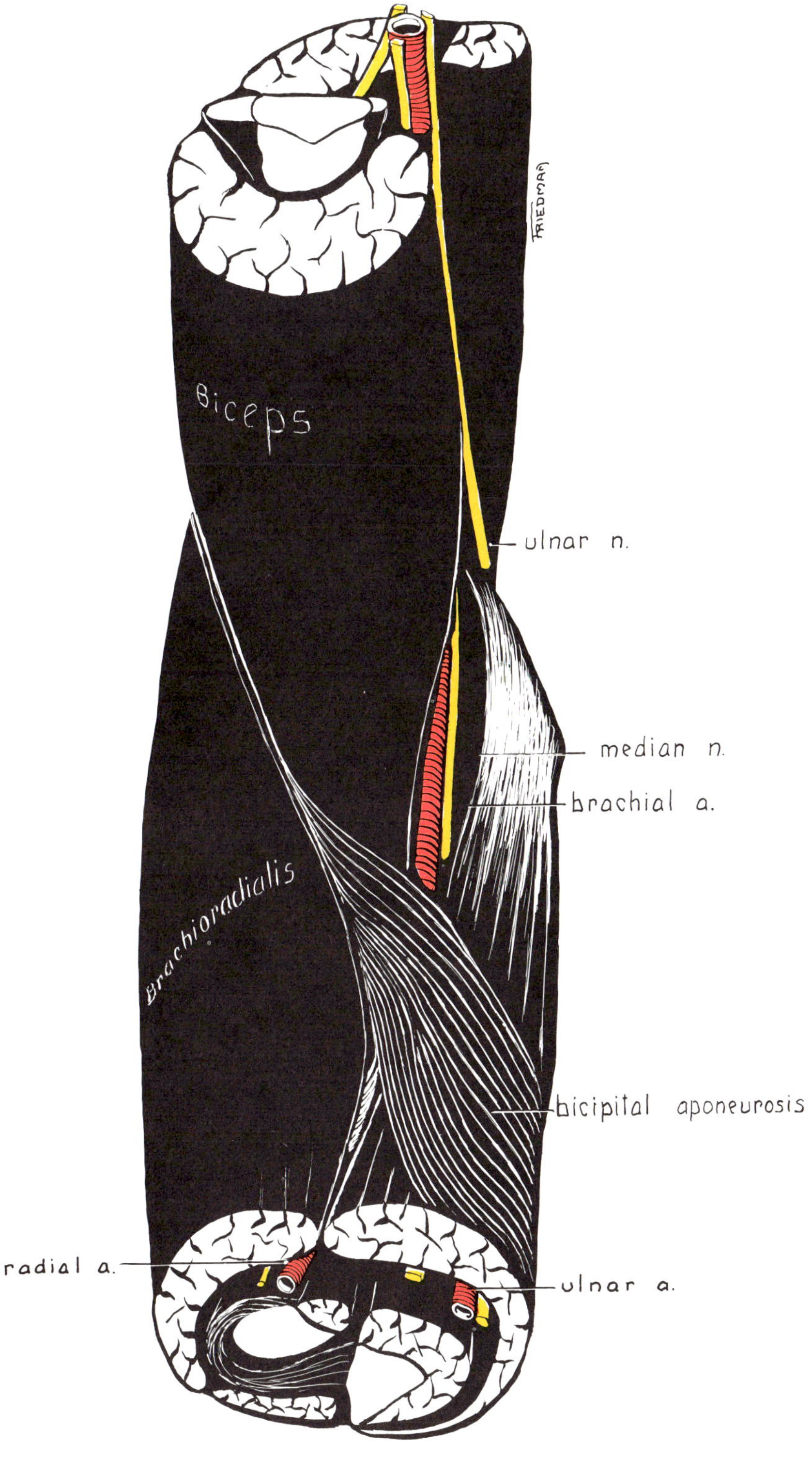
FRIEDMAN
Biceps
ulnar n.
median n.
brachial a.
Brachioradialis
bicipital aponeurosis
radial a.
ulnar a.

The Forearm and Hand

The Front of Forearm and Hand: I. The Radius and Ulna

The shaft of the **ulna** is roughly triangular in section for most of its length, becoming almost circular only at its lower end. Just one surface, the anterior, is seen here in its entirety, bounded by the sharp interosseous border laterally and a more rounded ill-defined (anterior) border on its medial side. The other two surfaces, posterior and medial, appeared in section in Figure 24.

The expansion of the ulna at its upper end forms the *coronoid process* and the *olecranon.* At its lower end, the ulna expands slightly to form a rounded *head* which articulates with the radius. Behind the head is the blunt *styloid process.*

The **radius** is named as if it and the ulna were lying head-to-toe. Thus, its *head* is superiorly placed. Below this is the cylindrical *neck* followed by the *tuberosity,* which receives the Biceps tendon posteriorly and a small bursa anteriorly. Below this again is the triangular *shaft.* The anterior surface of the shaft extends from the interosseous border medially to the *anterior oblique line* laterally. The lateral surface is indistinctly separated from the anterior, and at its lower end is prolonged as the *styloid process.* The lower end of the shaft is scooped out anteriorly. Medially, it presents a shallow concavity, the *ulnar notch,* to receive the head of the ulna.

The radius is hinged so that it can rotate about the ulna during pronation-supination through about 140 to 150 degrees while the ulna does not move at all. There are three parts to this hinge. The *superior radio-ulnar joint* involves the disclike head of the radius, which is held in place by the annular ligament against the shallow radial notch of the ulna. A similar *inferior radio-ulnar joint* involves the round head of the ulna and a shallow ulnar notch on the radius. A plate of fibrocartilage runs from the pit between the head and the styloid process of the ulna to the inferior margin of the ulnar notch and provides an inferior bearing surface. Between these upper and lower diarthrodial joints are the *interosseous membrane* and the *oblique cord.* The fibers of the interosseous membrane run downward and medially to the corresponding interosseous border of the ulna and so tend to prevent any upward movement of the radius on the ulna. Taken as a whole, the axis of rotation for the hinge is an oblique line passing through the middle of the head of the radius above and down through the middle of the head of the ulna below.

The radius at its lower end provides the whole of the bearing surface for the hand. It is, however, the ulna that forms the basic union with the humerus, so that the weight of a load carried in the hand passes first from hand to radius, then from radius to ulna, and ultimately from ulna to humerus. In the reverse direction, as in a fall on the outstretched hand, the force is transferred up from hand to radius, across by the interosseous membrane to ulna, and then to humerus. On occasion, the interosseous membrane may not be able to transfer fully such a force directed up the shaft of the radius, and in this case the head is driven against the capitulum and may be shattered.

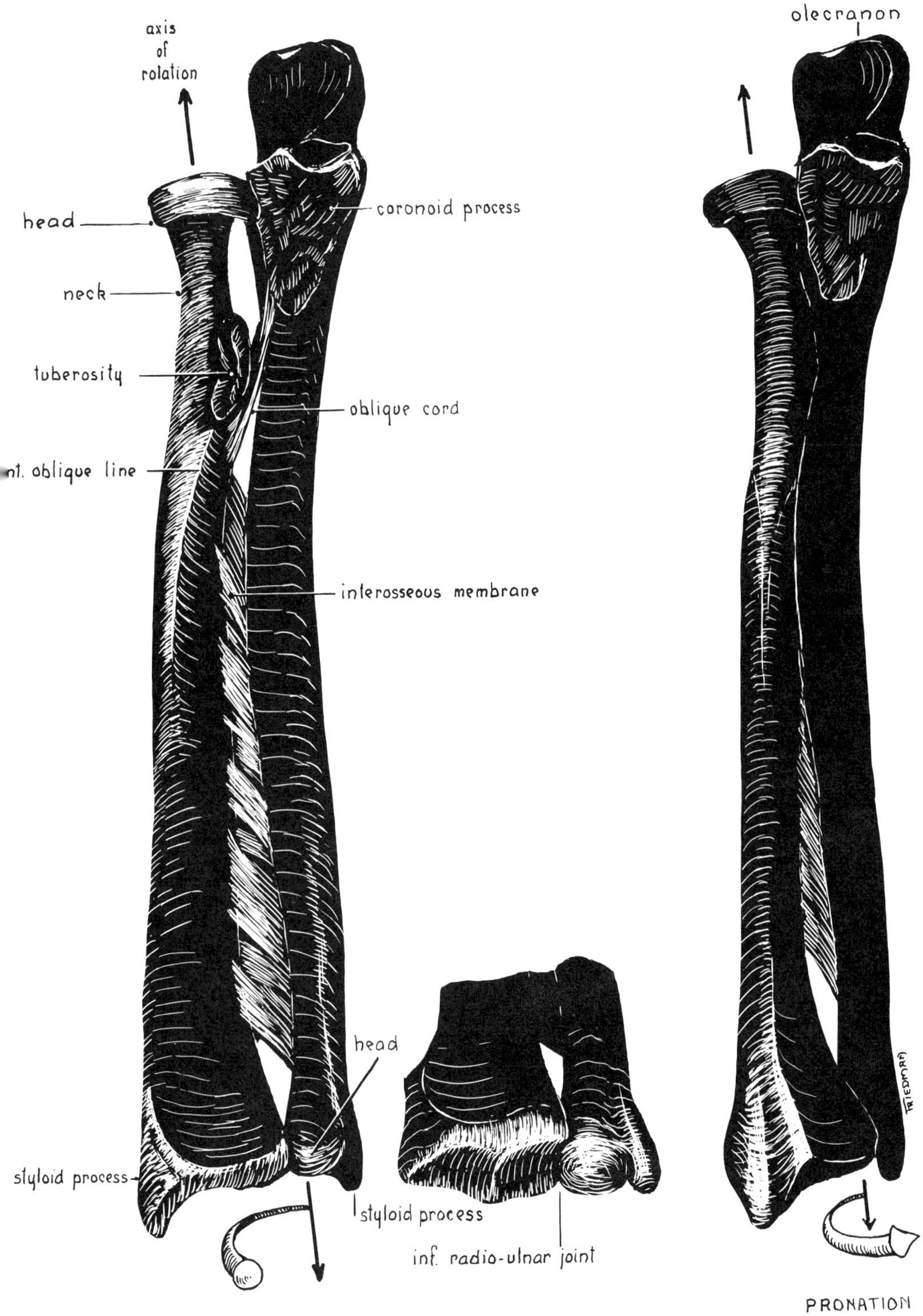
axis
of
rotation
olecranon
head
coronoid process
neck
tuberosity
oblique cord
nt. oblique line
interosseous membrane
head
styloid process
styloid process
inf. radio-ulnar joint
PRONATION

The Front of Forearm and Hand: II. The Supinators and Pronators

The muscles responsible for supination have already been described in connection with the skeleton of the elbow (Fig. 21). This movement is accomplished by a long and a short muscle. The long muscle is **Biceps,** which is, however, not really effective when the forearm is fully extended. It is most efficient when the forearm is at right angles to the arm, for, in this case, the tendon of Biceps acts at right angles to the long axis of the radius. The short muscle is **Supinator,** which has a superficial part arising from the lateral epicondyle of the humerus reinforced by a deep part which wraps around the upper half of the shaft of the radius from behind. This deep part of Supinator arises from the triangular fossa below the radial notch of the ulna.

Similarly, there is a long pronator and a short pronator. The long pronator is **Pronator Teres.** This is the most lateral of the superficial flexors that arise from the medial epicondyle of the humerus and owes its name to its ropelike appearance. It arises also from the supracondylar ridge immediately adjacent and, in addition, receives fibers from a deep head originating from the lower part of the coronoid process. Both heads converge to form a flat tendon of insertion which passes into the lateral surface of the radius a little below its midpoint. This muscle forms the medial limit of the cubital fossa and hence passes superficial to the tendon of insertion of Biceps and to the brachial and ulnar arteries. The median nerve passes downward between its two heads, innervating both as it does so.

The short pronator is **Pronator Quadratus.** It arises from the anterior aspect of the lower quarter of the shaft of the ulna and passes laterally into the anterior aspect of the corresponding portion of the radius as well as into a small triangular area on its medial aspect above the ulnar notch. It too is innervated by the median nerve through its anterior interosseous branch.

It is apparent that the pronator muscles do not have the bulk or mechanical advantage of the supinators, especially when the forearm is partially flexed and the Biceps is brought into play. It is for this reason that supination is a much more powerful movement than pronation. To take advantage of this for a predominantly right-handed population, screws and similar threaded items are made to be turned by supination of the right forearm.

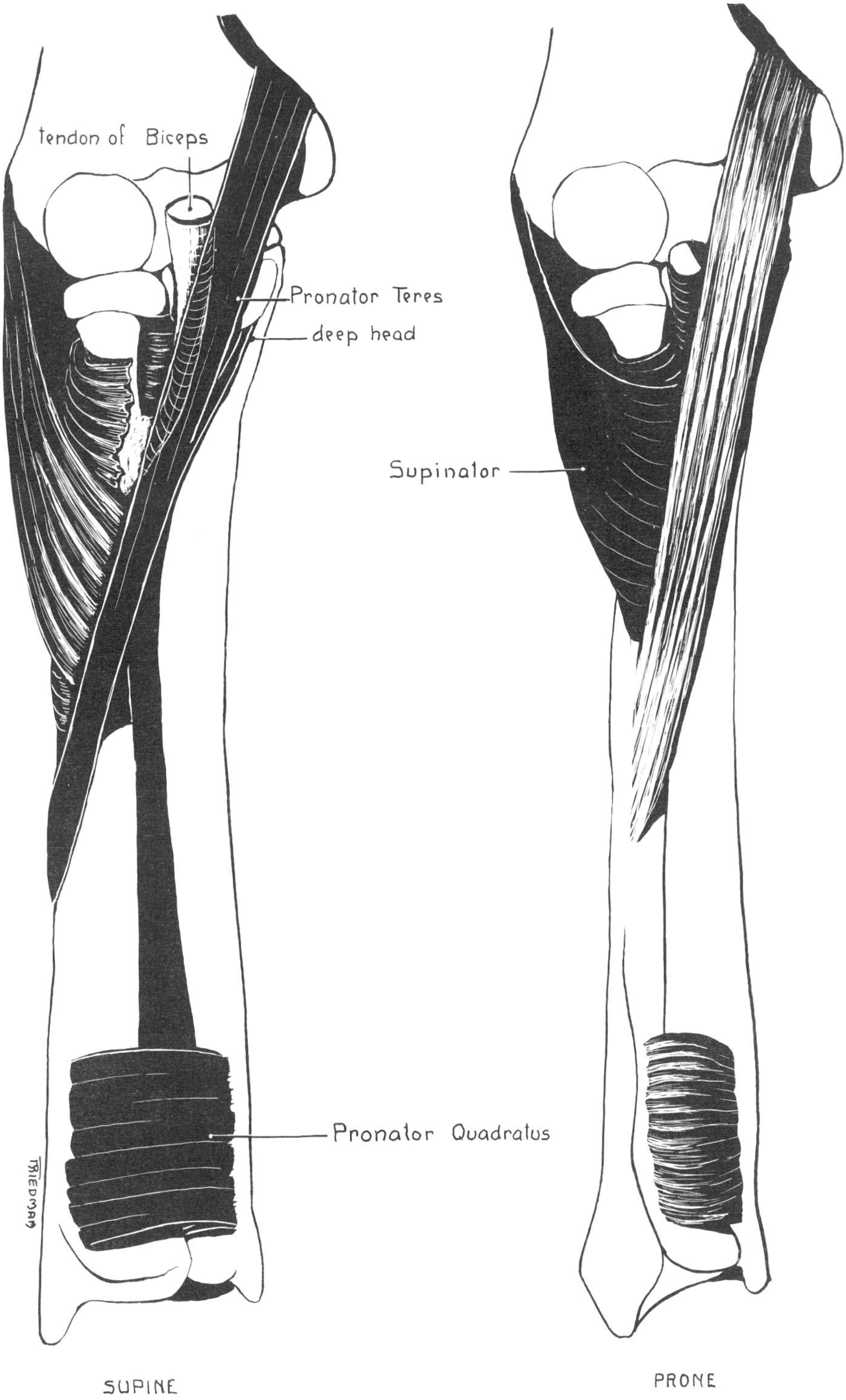
tendon of Biceps
Pronator Teres
deep head
Supinator
Pronator Quadratus
SUPINE
PRONE

The Front of Forearm and Hand: III. Flexion of the Wrist

Two rows of **carpal bones** unite the radius to the bones of the hand. Although these carpal bones are commonly termed wrist bones, it is worth remarking that in the living subject, at least, the distal bones form the proximal part of what we ordinarily consider the hand.

The proximal row of carpal bones consists of *scaphoid, lunate,* and *triquetral,* in that order passing from lateral to medial side, and the small *pisiform bone* which sits on the triquetral. The **proximal radiocarpal joint** involves the lower surface of the radius together with the cartilaginous disc that joins it to the ulna, meeting with the dome-shaped proximal surfaces of scaphoid and lunate, and a rim of the triquetral.

The distal row of carpal bones consists of *trapezium, trapezoid, capitate,* and *hamate* in that order passing from lateral to medial side. The **midcarpal joint** extends in S-shaped fashion between the two rows of carpal bones. This joint allows the bones of the carpus to accommodate themselves to fine movements of the fingers.

The distal row of carpal bones is united to the metacarpals in a series of individual joints. Of these **carpometacarpal joints** the first, between trapezium and the first metacarpal bone, is separate from the others. It provides a saddle-shaped surface on which the thumb moves freely. The second joint is a mortice into which the base of the second metacarpal bone fits so as to articulate mainly with the trapezoid while restrained by the trapezium laterally and the capitate medially. The capitate provides the major surface of articulation for the third metacarpal bone while the hamate joins with the fourth and fifth metacarpals. The carpometacarpal joints for the four fingers intercommunicate with one another, and there are cleftlike extensions between the metacarpals at their bases.

Any of the muscles that run across the carpus between forearm and hand obviously can produce flexion of the wrist in addition to their other actions. If they are placed toward the lateral side of the wrist, they also abduct the hand, and if on the medial side, they adduct the hand. Two muscles of the forearm, however, act only on the wrist and have no other function. Both arise from the common flexor origin and hence both belong to the superficial group of flexors. One, Flexor Carpi Radialis, has the most lateral tendon on the front of the wrist; the other, Flexor Carpi Ulnaris, the most medial. The tendon of **Flexor Carpi Radialis** passes down in front of the carpus and occupies a groove on the trapezium. It is inserted into the anterior aspect of the bases of both the second and third metacarpals. **Flexor Carpi Ulnaris** attaches to the pisiform bone and through the *pisometacarpal ligament* continues into the anterior aspect of the base of the fifth metacarpal.

Flexor Carpi Radialis, like the rest of the superficial flexor group, is innervated from its deep surface by the median nerve. Flexor Carpi Ulnaris, unlike the rest of this group, is innervated by the ulnar nerve which lies just deep to it.

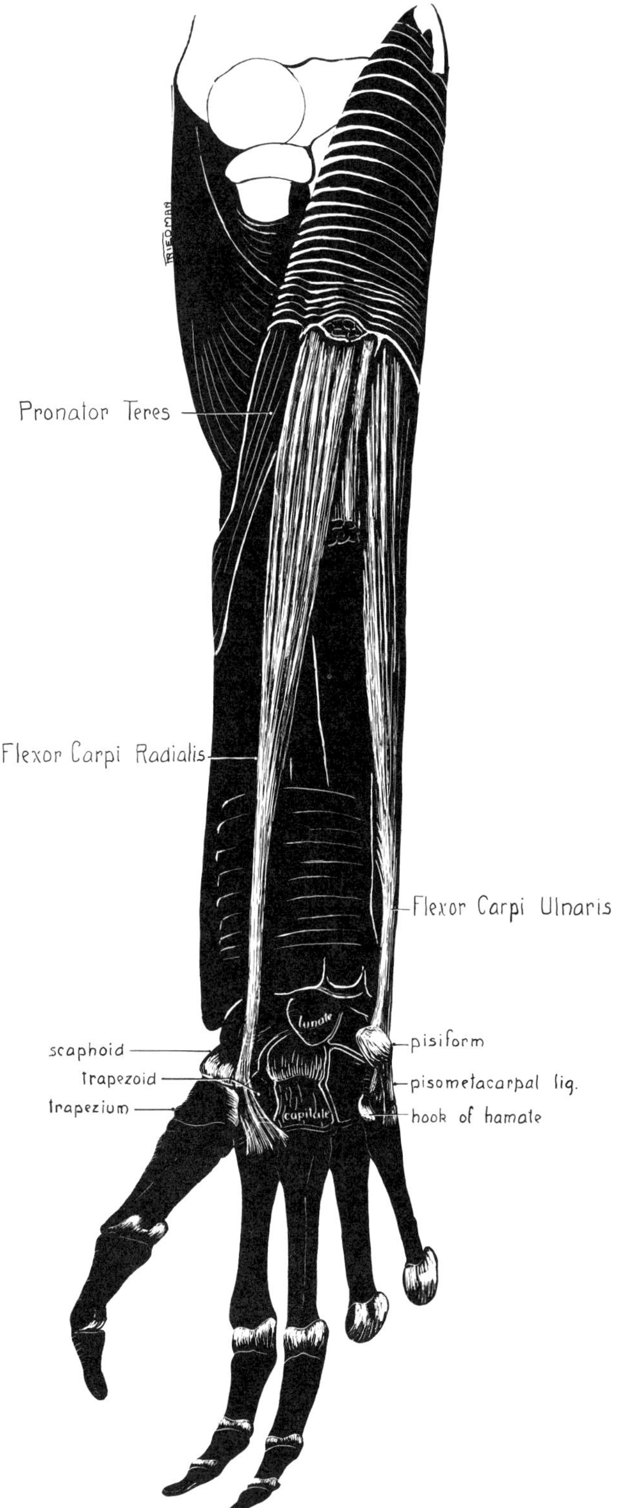
Pronator Teres
Flexor Carpi Radialis
Flexor Carpi Ulnaris
lunate
scaphoid
trapezoid
trapezium
capitate
pisiform
pisometacarpal lig.
hook of hamate

The Front of Forearm and Hand: IV. The Deep Flexor Muscles

The deep flexors of the fingers arise from as much of the surface of the radius and ulna as remains available to them (Fig. 26). In the case of the radius this means the anterior surface between the anterior oblique line and the insertion of Pronator Quadratus. In the case of the ulna this means the anterior aspect of the bone between the coronoid process above and Pronator Quadratus below as well as the adjacent part of the medial surface right up to the posterior border. In addition, the area of origin for these muscles is increased by extension onto the interosseous membrane.

The high degree of independence of the thumb is ensured by the complete separation of its deep flexor, Flexor Pollicis Longus, from the remainder of the deep group, the Flexor Digitorum Profundus. **Flexor Pollicis Longus** occupies the radial portion of the general area of origin for the deep flexors. Its tendon passes down in front of Pronator Quadratus to cross the insertion of Flexor Carpi Radialis and enter the thumb. Here it runs directly in front of the first metacarpal and the proximal phalanx of the thumb to the base of the distal phalanx. It is quite superficial as it lies against the proximal phalanx and can be readily palpated through the skin. As a result it is often accidentally cut here.

Flexor Digitorum Profundus arises from the ulnar portion of the area for deep flexors. It divides into four tendons which cross the carpus and run directly in front of their corresponding metacarpals. These tendons then proceed into osseofibrous tunnels anterior to the fingers and so to the anterior aspect of the distal phalanges where they insert.

Four **Lumbricals,** named for their wormlike appearance arise from the four tendons of Flexor Digitorum Profundus and run distally to the back of the fingers at the metacarpophalangeal joint. Each runs from front to back around the radial side of the respective finger, and inserts into the dorsal expansion (Fig. 42). Each of the first two Lumbricals originates from one tendon only. The third Lumbrical originates from the adjacent sides of the second and third tendons and the fourth from the third and fourth tendons.

Flexor Pollicis Longus and the portion of Flexor Digitorum Profundus passing to the second and third fingers are innervated by the median nerve through its branch, the *anterior interosseous nerve,* which descends on the interosseous membrane. The more medial portion of Flexor Digitorum Profundus, concerned with the function of the fourth and fifth fingers, receives its innervation, like the overlying Flexor Carpi Ulnaris, from the *ulnar nerve*. The Lumbricals are similarly innervated, the lateral one or two through the median nerve, the more medial two through the ulnar.

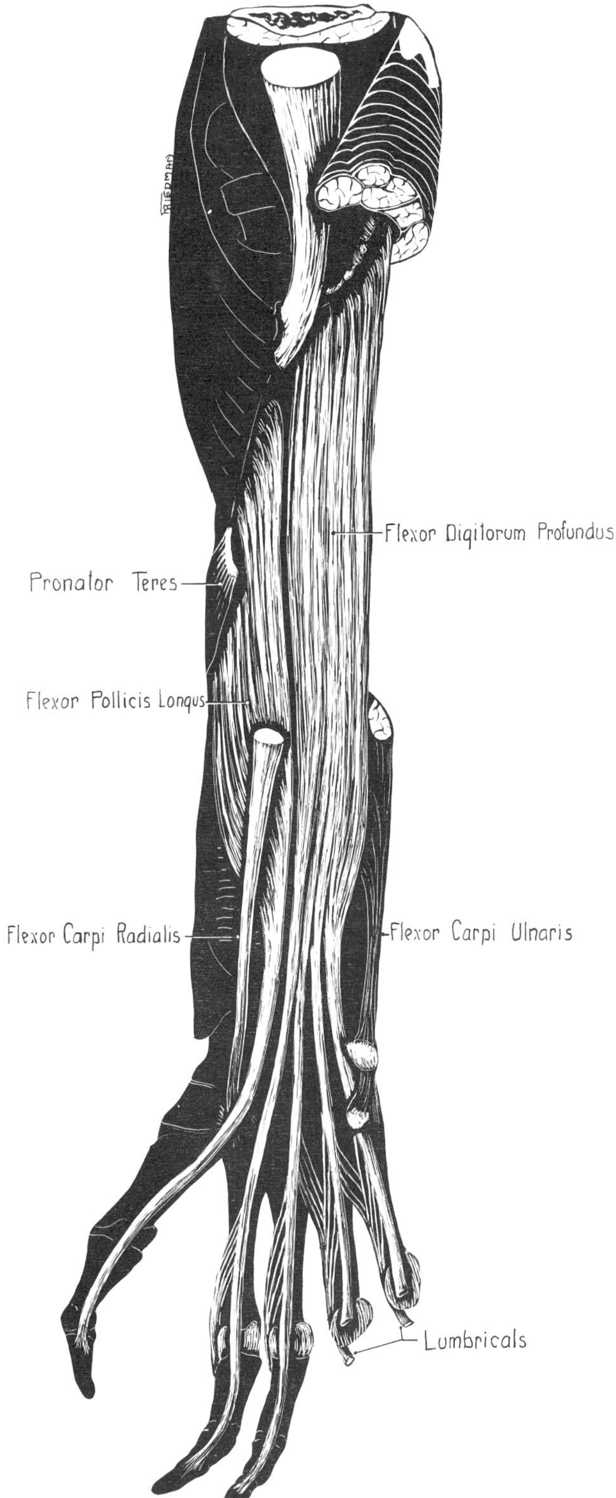
Flexor Digitorum Profundus
Pronator Teres
Flexor Pollicis Longus
Flexor Carpi Radialis
Flexor Carpi Ulnaris
Lumbricals

The Front of Forearm and Hand: V. The Radial and Ulnar Arteries

The **radial artery** arches gently lateralward from its point of origin in the cubital fossa, crossing in succession the tendon of Biceps and the Supinator muscle. It then continues downward, becoming progressively more superficial as the tapering muscles of the extensor group that overlap it withdraw from the front of the forearm. After leaving Supinator it lies briefly on the tendon of insertion of Pronator Teres, continues on Flexor Pollicis Longus, and then on Pronator Quadratus, finally reaching bone at the lower end of the radius. It then winds dorsally around the lateral margin of the carpus, i.e., around the scaphoid, and then behind the trapezium, but comes back again into the hand between the bases of the first two metacarpals. In the hand it runs across the deepest part of the palm as the *deep palmar arch.*

The radial artery is subcutaneous in the lower quarter of the forearm; and it is here, immediately lateral to the tendon of Flexor Carpi Radialis, that the pulse is most conveniently taken by compressing the artery against the resistance of the radius. The beat of the artery can also be appreciated, although less easily, as the vessel winds around the lateral aspect of the carpus.

The **ulnar artery** was seen earlier to pass medially across the upper part of Flexor Digitorum Profundus (Fig. 23). It then changes direction to run vertically down to the wrist, which it reaches at the lateral side of the pisiform bone. It makes a gentle arch across the palm superficial to all of the long tendons (some yet to be described) as the *superficial palmar arch.* Throughout almost the whole of its course in the forearm the ulnar artery is under cover of the superficial flexor muscles. As it approaches the pisiform bone it becomes quite superficial just lateral to the tendon of Flexor Carpi Ulnaris.

Both the radial and the ulnar arteries give off numerous branches to nourish muscles they pass. In addition, both vessels give *recurrent branches* which run back up to participate in the anastomoses around the elbow joint. The ulnar artery, in addition, gives off a constant large branch, the *common interosseous artery,* which passes toward the upper end of the interosseous membrane where it divides into a large *posterior interosseous artery* to nourish the muscles of the extensor compartment of the forearm and a smaller *anterior interosseous artery,* which descends on the interosseous membrane between Flexor Pollicis Longus and Flexor Digitorum Profundus. As it reaches the wrist, the radial artery frequently gives off a *superficial palmar branch* which runs across the front of the wrist to complete the superficial palmar arch of the ulnar artery. Similarly, the ulnar artery at the pisiform usually gives off a *deep palmar branch* which passes into the palm to complete the deep palmar arch from the radial artery.

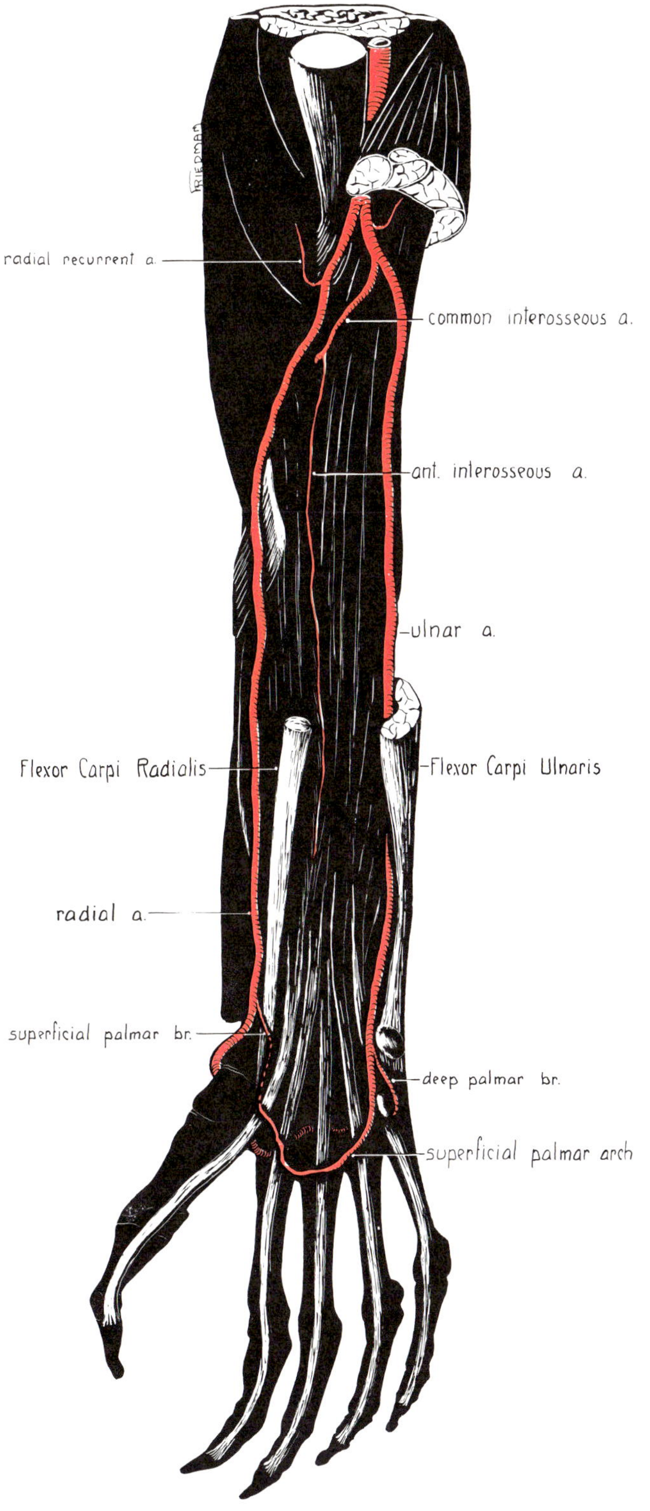
FRIEDMAN
radial recurrent a.
common interosseous a.
ant. interosseous a.
ulnar a.
Flexor Carpi Radialis
Flexor Carpi Ulnaris
radial a.
superficial palmar br.
deep palmar br.
superficial palmar arch

The Front of Forearm and Hand: VI. The Radial, Median, and Ulnar Nerves

The **radial nerve** enters the forearm lying on Supinator and covered over by Brachioradialis. Here it gives off branches to the two extensor muscles arising from the lateral supracondylar ridge, and then splits into a large deep and a small superficial branch. The *deep branch* passes from front to back of the forearm between the two heads of Supinator to innervate the remaining extensor muscles. The *superficial branch* is of lesser moment, since it is little more than a cutaneous nerve carrying sensory fibers from the skin of the lateral side of the dorsum of the hand. It runs vertically downward to reach the lateral side of the radial artery, which it accompanies to the lower part of the forearm. Here it leaves the artery to wind posteriorly around the radius and breaks up into branches for the dorsum of the hand and for the lateral two or three fingers.

The **median nerve** enters the forearm lying on Brachialis at the medial side of the brachial artery. It passes almost vertically downward, inclining laterally enough to reach the midline of the forearm. As it does so, it crosses the ulnar artery although separated from it by the deep head of Pronator Teres. It lies for most of its course immediately deep to the superficial flexor group, specifically to Flexor Digitorum Superficialis, and is clamped by fascia to the undersurface of this muscle. Just above the wrist, the taper of its covering muscle leaves it almost superficial at the medial side of the tendon of Flexor Carpi Radialis. It then passes into the palm where we shall see it split into its terminal branches. These are crossed by the superficial palmar arch.

The **ulnar nerve** enters the forearm from behind the medial epicondyle and is at once covered by the superficial flexor group of muscles as these arise from the common flexor origin—specifically, at first, by the upper edge of the aponeurosis of Flexor Carpi Ulnaris (Fig. 39). It stays in this position and reaches the medial side of the ulnar artery with which it runs down the forearm to enter the hand. The ulnar nerve supplies Flexor Carpi Ulnaris and the medial portion of Flexor Digitorum Profundus by branches given off just below the medial epicondyle. The sharp, free upper margin of the aponeurosis may, in bedridden patients, constrict the ulnar nerve sufficiently to cause temporary paralysis resulting in "ulnar claw hand."

The other muscles of the flexor compartment of the forearm—i.e., Pronator Teres, Flexor Carpi Radialis, Flexor Digitorum Superficialis, Flexor Pollicis Longus, and the lateral part of Flexor Digitorum Profundus—are all supplied by the median nerve directly or through the *anterior interosseous nerve.* Since Pronator Quadratus is also innervated by the anterior interosseous nerve, it is clear that the function of pronation requires only an intact median nerve. Supination, by contrast, is accomplished with two nerves—the musculocutaneous to Biceps and the radial to Supinator.

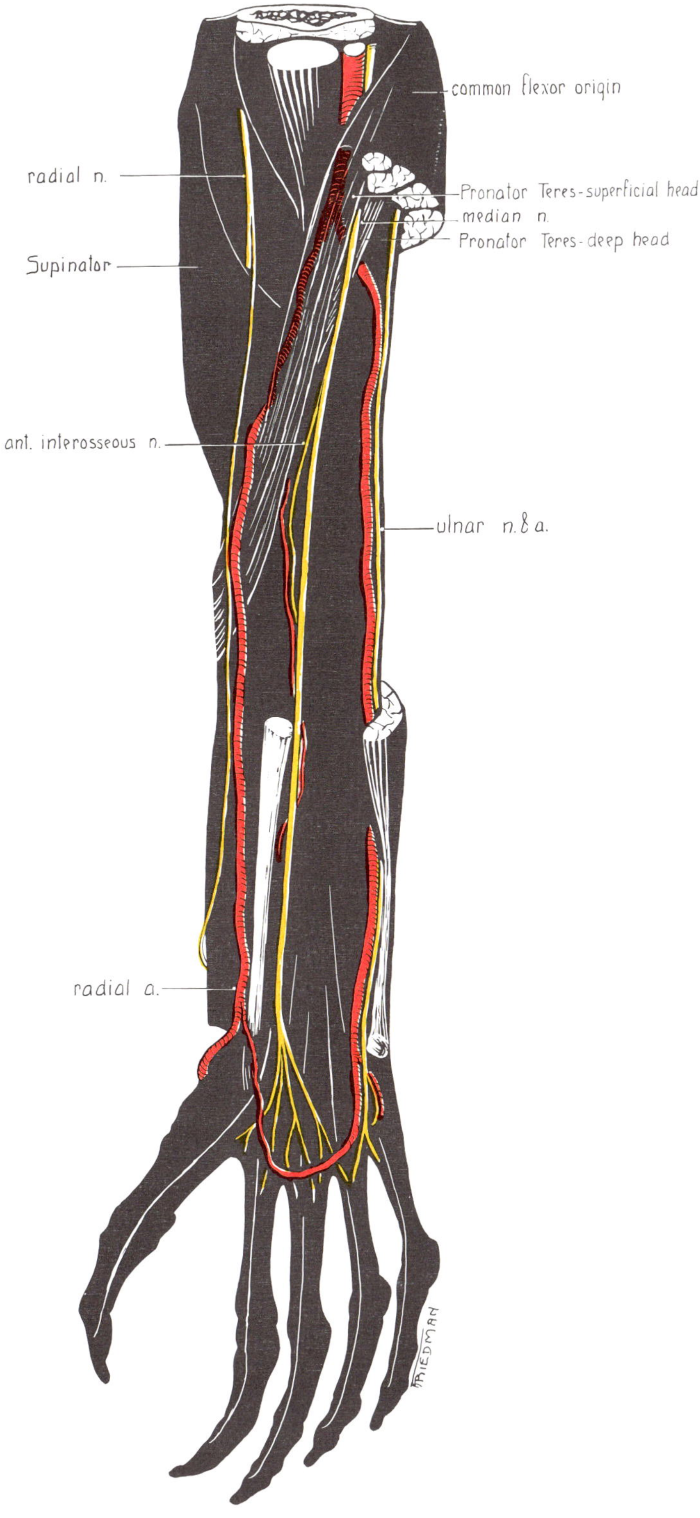
common flexor origin
radial n.
Pronator Teres-superficial head
median n.
Pronator Teres-deep head
Supinator
ant. interosseous n.
ulnar n.& a.
radial a.
FRIEDMAN

The Front of Forearm and Hand: VII. The Superficial Flexors of the Fingers

Flexor Digitorum Superficialis directly overlies the deep flexors, Flexor Pollicis Longus and Flexor Digitorum Profundus. It arises broadly from the common flexor origin on the medial epicondyle, from the dense fascia between the various muscles of the superficial flexor group, and, as a flat radial head, from the anterior oblique line of the radius. It is the most deeply placed of the superficial flexor group and is seen in its full extent only if Pronator Teres and Flexor Carpi Radialis are removed. From the broad muscle belly, tendons to the four fingers arise and are bunched together at the wrist, the tendons to the third and fourth fingers concealing those to the second and fifth. The tendons pass across the palm to enter the osseofibrous flexor tunnel of the digits, where each one splits in two to form an archway for the tendon of Profundus. The two slips wrap around the tendon to insert into the anterior aspect of the middle phalanx of the finger. Flexor Digitorum Superficialis is innervated by the median nerve, which clings to its deep surface throughout most of its course.

At the wrist, the **median nerve** escapes from under cover of Flexor Digitorum Superficialis to lie between it and Flexor Carpi Radialis, and here it is quite superficial. From this position it passes forward into the palm, where it splits into *digital nerves* which supply the skin of the lateral three and a half digits. The **ulnar nerve** correspondingly innervates the skin covering the medial one and a half digits. There is considerable variation in the area of the hand and fingers supplied by median and ulnar nerves. In all cases, however, these nerves supply the fronts and sides of the fingers and, in addition, carry around to the back of the digit at about the middle phalanx to innervate the dorsal aspect of the terminal portion of the digit and the nail bed. All of the digital branches of the ulnar and median nerves lie superficial to the long flexor tendons and are crossed by the superficial palmar arch.

Perhaps the most important single branch given off by the median nerve in the palm is a *recurrent branch,* which arches laterally and slightly backward at the base of the thumb to enter the three thenar (ball of thumb) muscles which it then supplies. These muscles—Abductor Pollicis Brevis, Flexor Pollicis Brevis and Opponens Pollicis (described in Figs. 35 and 36)—can be completely paralyzed by a small and not particularly deep wound in the proximal part of the palm if this is placed with unfortunate accuracy.

At the wrist both deep and superficial tendons of the fingers are enclosed in a common synovial sheath often termed the *ulnar bursa.* This sheath extends 1 or 2 cm down into the palm and often up a similar distance above the carpus. It commonly continues as the synovial sheath enclosing the deep and superficial tendons going to the little finger. The synovial sheaths similarly surround the pairs of tendons as they run along each of the fingers, but these sheaths, except for that of the little finger, are usually separate from the ulnar bursa. Flexor Pollicis Longus is enclosed within a synovial sheath unique to it. This sheath, clinically termed *radial bursa,* extends from just above the wrist distally the whole length of the thumb.

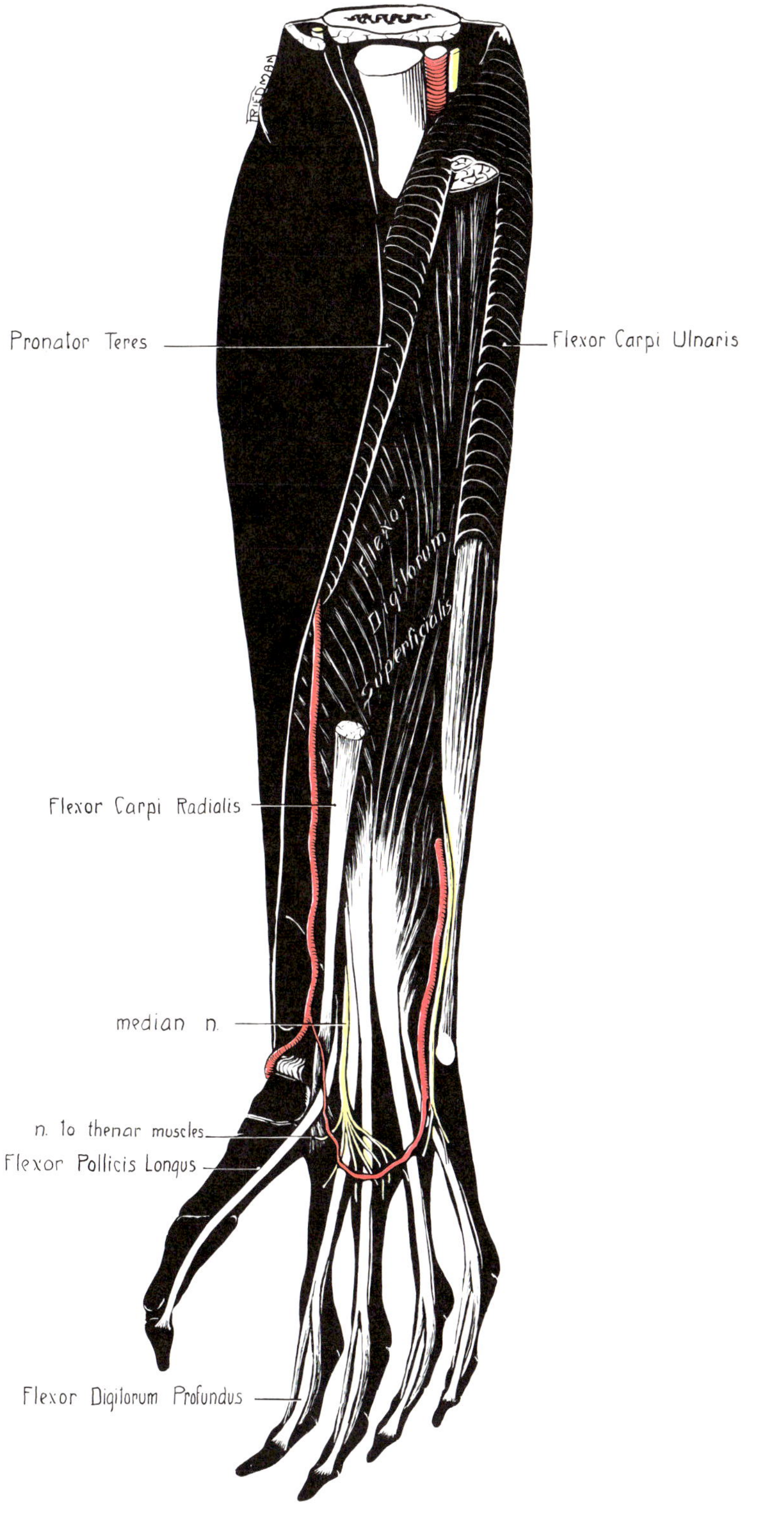

Pronator Teres
Flexor Carpi Ulnaris
Flexor Digitorum Superficialis
Flexor Carpi Radialis
median n.
n. to thenar muscles
Flexor Pollicis Longus
Flexor Digitorum Profundus

The Front of Forearm and Hand: VIII. The Flexor Retinaculum

All of the structures that pass down from the forearm to the front of the palm run, with only few exceptions, deep to the flexor retinaculum, a stout ligamentous band across the front of the carpus. The **flexor retinaculum** arises from the most prominent bony points on the anteromedial and anterolateral sides of the carpus. Medially these high points are the pisiform and below that the hook of the hamate. On the lateral side the high points are the tubercle of the scaphoid and the so-called ridge of the trapezium.

Palmaris Longus is a small muscle (absent in about 20 per cent of subjects) which arises from the common flexor origin. It has only a small muscular belly at the upper part of the forearm and a long narrow tendon which descends almost directly in the middle of the flexor aspect of the forearm. This tendon blends with the flexor retinaculum and passes into the palm to blend with and pull on the palmar aponeurosis. It can be made to stand out by actively cupping the palm.

The space between the flexor retinaculum and the concave anterior aspect of the carpus is termed the **carpal canal.** At the posterolateral extremity of the canal, Flexor Carpi Radialis, held by retinacular fibers in its groove in the trapezium, passes directly to its insertion into the bases of the second and third metacarpals. Next alongside is the row of deep flexor tendons beginning laterally with the tendon of Flexor Pollicis Longus and continuing with those of Flexor Digitorum Profundus. On the next plane anterior to these tendons, are the four tendons of Flexor Digitorum Superficialis, the two destined for the second and fifth fingers lying deep to those for the third and fourth fingers. Immediately under cover of the flexor retinaculum, and just lateral to the tendons of Flexor Digitorum Superficialis, is the stout median nerve passing into the palm. Its recurrent branch to the muscles of the thenar eminence arises just as the nerve leaves the carpal canal to enter the palm. The median nerve is often compressed in the tightly packed carpal canal; the ensuing palsy is known as "carpal tunnel syndrome."

The ulnar nerve and artery are the only structures of consequence entering the palm superficial to the flexor retinaculum; the small *palmar branch of the radial artery* also lies here.

All of the muscles making up the superficial group of flexors that arise from the common origin on or immediately above the medial epicondyle have now been described. From lateral to medial they are Pronator Teres which extends only to the midpoint of the radius, Flexor Carpi Radialis which is the most lateral of the tendons of the front of the wrist, Flexor Digitorum Superficialis with the inconstant Palmaris Longus superficial to it, and most medial, Flexor Carpi Ulnaris. The radial artery lies immediately lateral to Flexor Carpi Radialis, while the ulnar artery lies directly lateral to Flexor Carpi Ulnaris.

Both the superficial and the deep tendons of the fingers are conducted to their termination through an osseofibrous digital tunnel. The fibrous capsules of the metacarpophalangeal joints are tied to one another by the *deep transverse metacarpal ligaments.* There is thus only little movement of the metacarpals permitted. The fingers, of course, are entirely free.

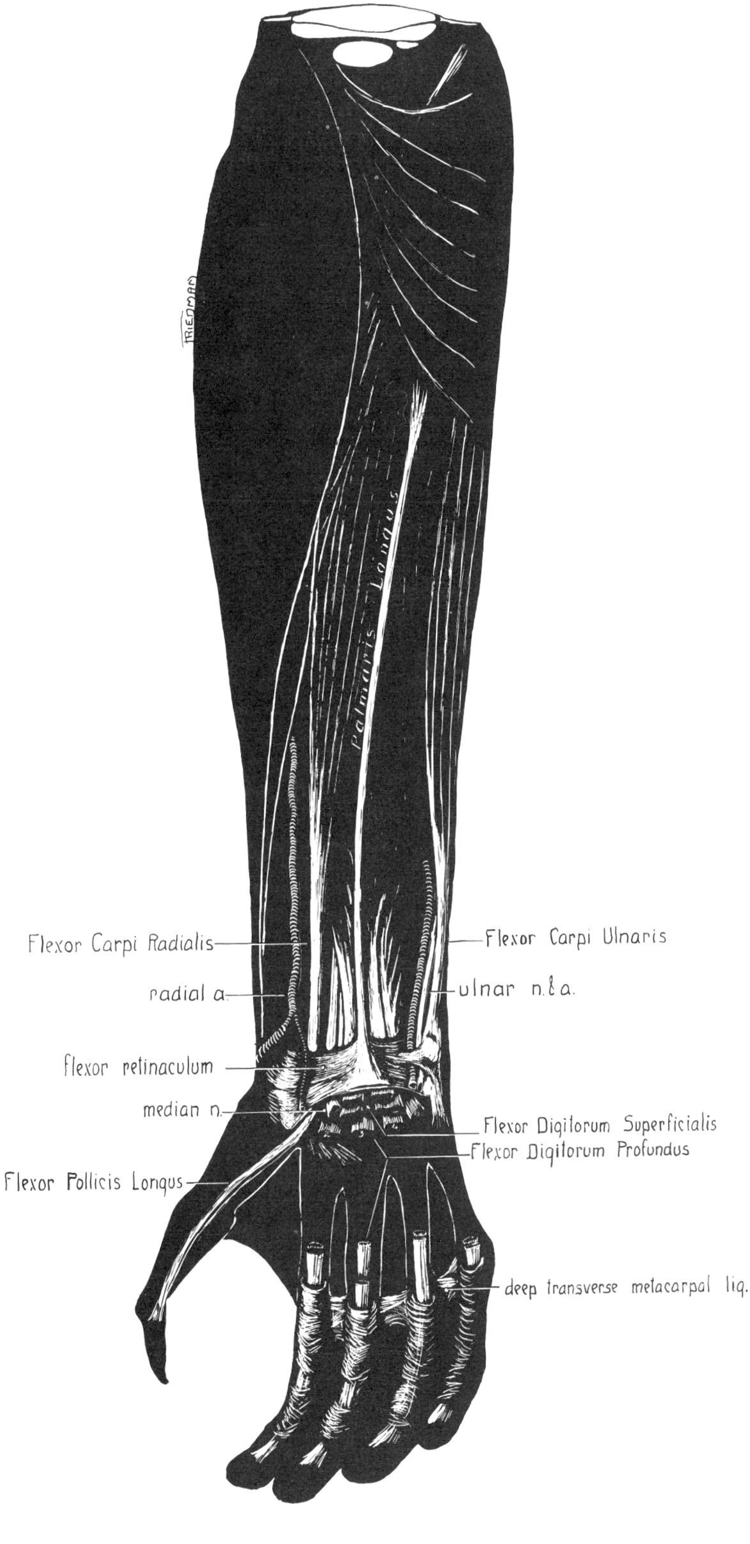
FRIEDMAN
Palmaris Longus
Flexor Carpi Radialis
radial a.
flexor retinaculum
median n.
Flexor Pollicis Longus
Flexor Carpi Ulnaris
ulnar n. & a.
Flexor Digitorum Superficialis
Flexor Digitorum Profundus
deep transverse metacarpal lig.

The Hand

The Intrinsic Muscles:
I. The Dorsal Interossei

The side-to-side movement of the fingers is described by reference to the middle finger. Abduction is thus the movement of the thumb and index finger on one side and the fourth and fifth fingers on the other away from the midline of the hand which passes through the third digit. In this sense, since, as simple inspection of the fingernail shows, the thumb lies at right angles to the other fingers, its abduction is at right angles to abduction of the other fingers. Adduction of the fingers is defined as the movement of the fingers toward the midline, and adduction of the thumb is the same movement carried out at right angles to that of the fingers.

Abduction of the fingers is produced by the Dorsal Interossei muscles. Four **Dorsal Interossei** are present, each arising from the sides of the metacarpal bones adjacent to the four intermetacarpal clefts. The two lateral Interossei run around the lateral side of the second and third fingers respectively, passing behind the deep transverse metacarpal ligaments, and insert primarily into the extensor (dorsal) expansion which overlies the proximal phalanx of each of the fingers as the "extensor hood" (Fig. 42). The two medial Dorsal Interossei arise similarly but pass around the ulnar side of the metacarpophalangeal joint of the third and fourth digits.

Since there is a Dorsal Interosseus acting on each side of the metacarpophalangeal joint of the middle finger, this digit can be waggled from side to side without the help of other muscles. The first Dorsal Interosseus moves the index finger away from the middle finger; the fourth Dorsal Interosseus moves the fourth finger away from the middle finger. Abductor Pollicis Brevis produces this same effect for the thumb and Abductor Digiti Minimi for the little finger (Fig. 36).

All Dorsal Interossei are innervated by the deep branch of the ulnar nerve.

The Dorsal Interossei **FIGURE 33**

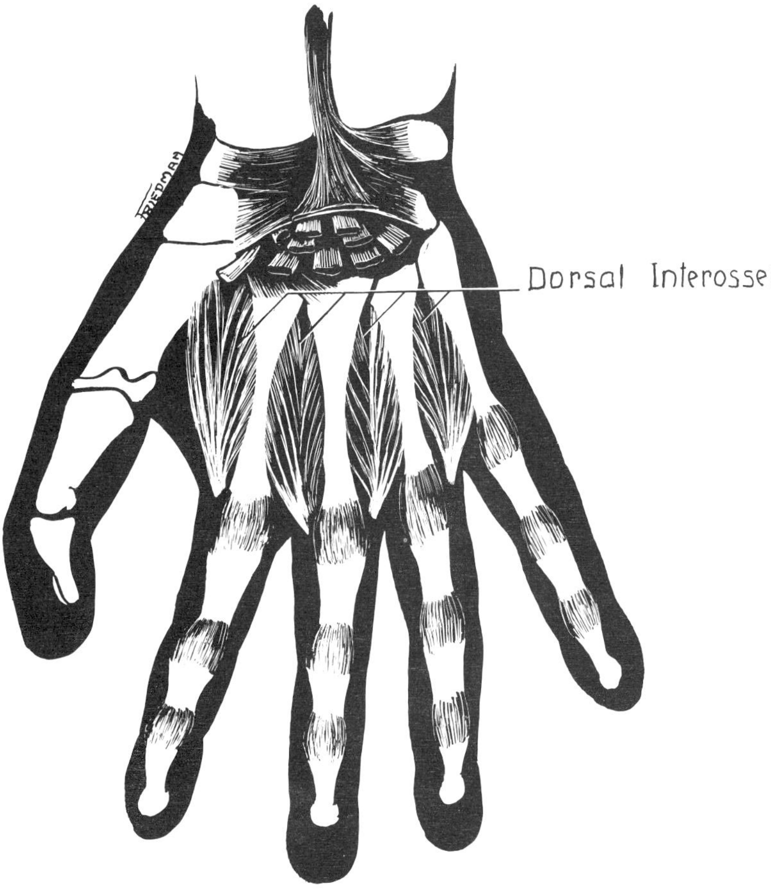

The Intrinsic Muscles: II. The Palmar Interossei

The Palmar Interossei, also innervated by the deep branch of the ulnar nerve, draw the fingers together and so cup the palm. As simple examination of the living hand will show, far less strength is required for the fingers to fall naturally together as in cupping than for them to be drawn apart. The Palmar Interossei, as shown here in a cupped hand, are thus very small indeed, the first one being rudimentary.

The first and second **Palmar Interossei** are placed so as to bring the thumb and index finger toward the midline. The first thus arises from the medial aspect of the base of the first metacarpal, the second arises from the medial aspect of the shaft of the second metacarpal, and the tendons of both pass back to the extensor expansion around the medial side of the respective fingers. Similarly, to draw the fourth and fifth fingers in toward the midline, the fourth and fifth Palmar Interossei arise from the lateral side of the shafts of the fourth and fifth metacarpals and pass by tendons around the lateral side of the respective metacarpophalangeal joints into the extensor expansion. No Palmar Interossei are required for the third digit, since its side-to-side movement is fully provided for by the Dorsal Interossei.

The fundamental musculoskeletal constituents of the palm of the hand have now been examined. Proximally, the carpal canal, formed between the anterior aspect of the concavity of the carpal bones and the flexor retinaculum, contains the long tendons which run from forearm to hand, and the median nerve. The carpal canal is located at the base of the palm and extends for about 3 cm below the distal skin crease of the wrist. Beyond this, the palm includes the shafts of the metacarpal bones, with Dorsal and Palmar Interossei filling up the clefts between them, and extends up in front of the bases of the proximal phalanges of the fingers.

The radial artery winds around the lateral aspect of the carpus and contributes to the arterial network on the dorsal aspect of these bones. It then passes between the two heads of the first Dorsal Interosseus to enter the palm as the **deep palmar arch.** The arch runs in a rather transverse fashion across the anterior aspect of the bases of the metacarpals.

The first branches of the deep palmar arch are to the thumb and index finger —the *princeps pollicis artery* and the *radialis indicis artery.* The princeps pollicis, or chief artery of the thumb, splits as it approaches the metacarpophalangeal joint and supplies both sides of the digit. The deep arch then gives off *palmar metacarpal arteries* which pass down each of the other interspaces. These link up with the corresponding larger *palmar digital arteries* from the superficial arch close to the metacarpophalangeal joints. Beyond this point of union, the vessels split to form *digital arteries,* which pass distally on the adjacent sides of the fingers. The medial side of the little finger is usually separately supplied from the superficial arch or, on occasion, from the deep arch.

Branches from the deep palmar arch pass upward to the anterior aspect of the carpus. The deep arch is usually completed by the *deep palmar branch of the ulnar artery.*

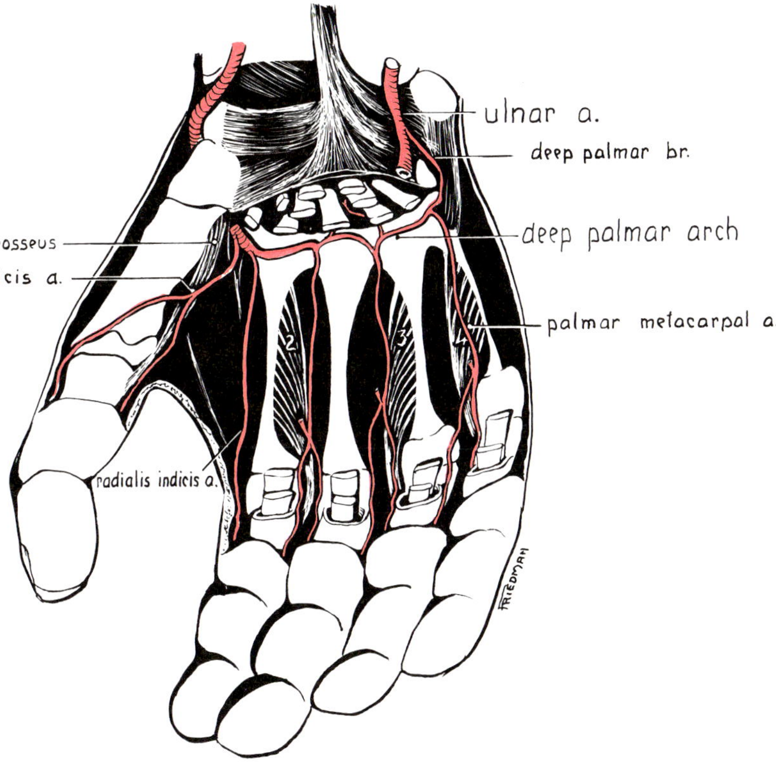
ulnar a.
deep palmar br.
deep palmar arch
1st Palmar Interosseus
princeps pollicis a.
palmar metacarpal a.
radialis indicis a.
2
3
4
FRIEDMAN

The Intrinsic Muscles:
III. The Deep Muscles of the Thumb and Little Finger

One intrinsic muscle of the palm is uniquely concerned with adduction of the thumb. In a sense, this muscle has taken over the function of the first Palmar Interosseus and by extending the attachment across the palm to the middle of the third digit has enormously magnified the effect. **Adductor Pollicis** arises by two heads which cover almost the entire central portion of the palm. The oblique head arises proximally from the central carpal bones and the adjacent parts of the bases of the metacarpals. The transverse head arises from the central ridge along the anterior aspect of the third metacarpal. From these origins the muscle tapers to a tendon which inserts primarily into the ulnar side of the base of the first phalanx of the thumb. The muscle may extend by a small tendinous slip to the radial side of the same phalanx. Like the Interossei, Adductor Pollicis is innervated by the *deep branch of the ulnar nerve.*

Since this muscle does all that the first Palmar Interosseus does, and considerably more, the latter is relatively insignificant. Adductor Pollicis not only draws the thumb toward the index finger but carries it across the palm until it is almost directly over the third digit. It is enclosed by its own dense fascial sheath, and this is complete enough to be considered a fascial compartment of the palm in its own right. It is worth noting that the upper portion of the muscle can be grasped between two fingers in the web of the first interspace.

The entrance of the radial artery and its branches to thumb and index finger is covered by Adductor Pollicis. Strictly speaking, the deep palmar arch begins only when the radial artery passes into the palm proper between the two heads of Adductor Pollicis.

The deepest of the thenar (ball of thumb) and of the hypothenar (ball of little finger) muscles are the Opponens Pollicis and the Opponens Digiti Minimi. Opponens Pollicis arises from the carpal bones that support the lateral end of the flexor retinaculum and from the adjacent part of the retinaculum itself. It passes as a flat sheet of muscle into the anterior aspect of the shaft of the first metacarpal. Like the other thenar muscles it is innervated by the *recurrent branch of the median nerve* which arises as the nerve enters the palm. In similar fashion, **Opponens Digiti Minimi** arises from the bones that support the medial end of the flexor retinaculum and from the adjacent part of the retinaculum itself. It passes down as a flat sheet of muscle which inserts into the anteromedial aspect of the shaft of the fifth metacarpal. Like the other hypothenar muscles it is innervated by the *deep branch of the ulnar nerve.*

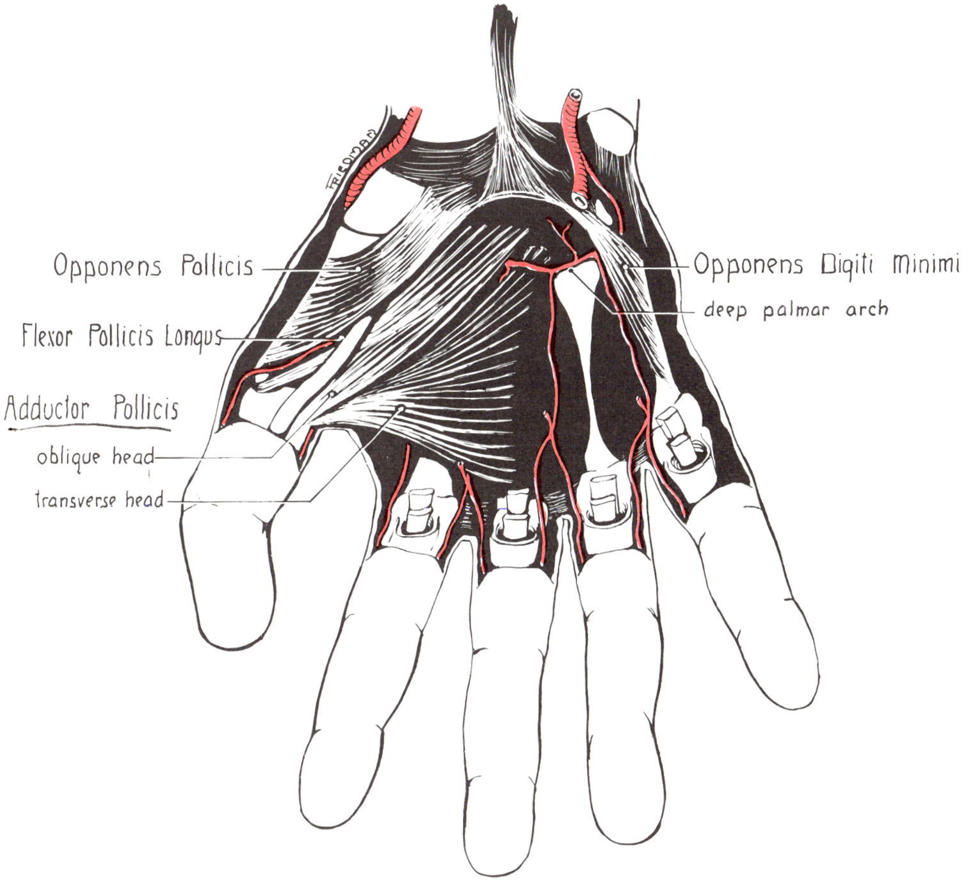
Opponens Pollicis
Flexor Pollicis Longus
Adductor Pollicis
oblique head
transverse head
Opponens Digiti minimi
deep palmar arch

The Intrinsic Muscles:
IV. The Superficial Muscles of the Thumb and Little Finger

A pair of superficial muscles completes the fullness of the thenar eminence. These arise as a flat sheet from the lateral end of the flexor retinaculum and its supporting carpal bones. They pass together to insert by a common tendon into the radial side of the base of the first phalanx of the thumb. **Abductor Pollicis Brevis** is the more lateral of the two and is only indistinctly separated from the medially placed **Flexor Pollicis Brevis.** The pull of these muscles on the radial side of the base of the first phalanx is counterbalanced by the pull of Adductor Pollicis on the ulnar side so that both units acting together produce quite powerful flexion at the metacarpophalangeal joint. A sesamoid bone is usually present at the insertion of each of these two units. The tendon of Flexor Pollicis Longus passes between them and is here quite superficial and vulnerable to injury.

The three thenar muscles—Abductor Pollicis Brevis, Flexor Pollicis Brevis, and Opponens Pollicis—are all innervated by the *recurrent branch of the median nerve.* They are encased in their own fascial enclosure which constitutes an independent fascial compartment in the hand. We have already identified another fascial compartment which houses the Adductor Pollicis alone. The large *midpalmar space* in front of Adductor Pollicis is also fascia-enclosed and constitutes a central palmar compartment which is further subdivided into lateral and medial portions. The Interossei between metacarpals are similarly enclosed in separate fascial compartments which communicate with the dorsum of the fingers alongside the tendons.

As in the case of the thumb, the little finger is moved by an abductor and flexor which arise similarly from the medial portion of the flexor retinaculum and the adjacent carpal bones. They are inserted in common on the ulnar side of the base of the first phalanx of the little finger. **Abductor Digiti Minimi** is the more medial and **Flexor Digiti Minimi** the more lateral of the two muscles. Together with the underlying **Opponens Digiti Minimi** they constitute the hypothenar group and are responsible for the fullness of the hypothenar eminence on the medial side of the palm. They are innervated by the *deep branch of the ulnar nerve,* which passes first between Flexor and Abductor Digiti Minimi and then between these muscles and Opponens on its way into the palm. They are all encased together in their own fibrous sheath which constitutes an independent fascial compartment in the hand.

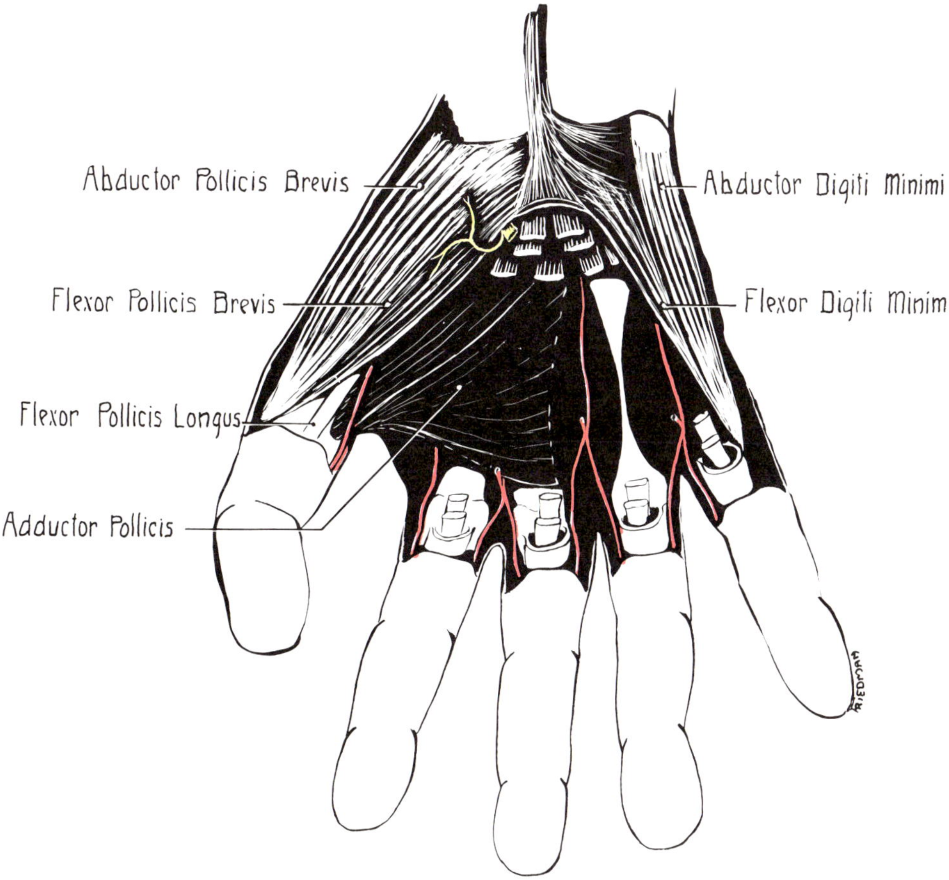
Abductor Pollicis Brevis
Flexor Pollicis Brevis
Flexor Pollicis Longus
Adductor Pollicis
Abductor Digiti Minimi
Flexor Digiti Minimi
RIEDMAN

The Digital Arteries and Nerves

The superficial and deep flexors of the fingers enter the palm from the carpal canal in a close bunching. They diverge from one another in pairs to fill the midpalmar space and enter the digits distally to be conveyed in osseofibrous tunnels to the middle and distal phalanges. They are overlain by the superficial palmar arch as well as by the digital branches of the ulnar and median nerves. In the palm, the arch and its branches are superficial to the nerves; this relationship is reversed at the sides of the digits.

The **ulnar artery,** with the **ulnar nerve** on its medial side, enters the palm lying in an incomplete retinacular compartment which holds the two structures against the lateral side of the pisiform bone and anterior to the flexor retinaculum proper. Here, just before entering the palm, the artery and nerve each give off a deep branch which runs to the floor of the palm between the hypothenar muscles. The *deep palmar branch of the ulnar artery* joins the radial to complete the deep palmar arch (Fig. 34). The *deep branch of the ulnar nerve,* a most important structure, supplies the motor innervation for the three hypothenar muscles, all of the Interossei, Palmar as well as Dorsal, Adductor Pollicis, and the medial two and occasionally three Lumbricals. It is thus entirely responsible for all of the intrinsic musculature of the hand exclusive of the thenar muscles and one or two Lumbricals.

Proceeding into the palm, the ulnar artery forms a gentle curve with its convexity downward, the *superficial palmar arch,* and this is usually completed by the *superficial palmar branch of the radial artery.* From this arch a series of *palmar digital arteries* runs down toward the fingers. At or near the webs of the fingers these join the palmar metacarpal arteries from the deep arch and, splitting, run as digital arteries along the sides of the fingers. The *digital arteries* run up to the tips of the fingers together with the digital nerves along the sides of the osseofibrous tunnel for the flexor tendons. They break up at the terminal phalanx into an extensive capillary network. The thumb is separately supplied directly from the radial artery as it enters the palm.

The *digital nerves* to the little finger and half of the fourth finger (approximately) arise as branches of the ulnar nerve in the palm. Those that supply the lateral three and a half fingers arise similarly from the median nerve. In each case, the digital nerve supplies sensory fibers to the anterior surface of the proximal half and all parts of the distal half of each finger. Before breaking up into these digital nerves, the median nerve gives off its *recurrent branch,* which supplies the motor innervation for the three thenar muscles.

In most regions of the body, cutaneous innervation is extensive and overlapping, and its interruption in limited areas produces no great disability. Not so in the case of the hand. Here, despite rich overlapping, the sense of touch is of such cardinal importance that even a seemingly minor nerve interruption may be significant.

The digital nerves can be interrupted by an appropriate anesthetic solution as they run up the sides of the fingers.

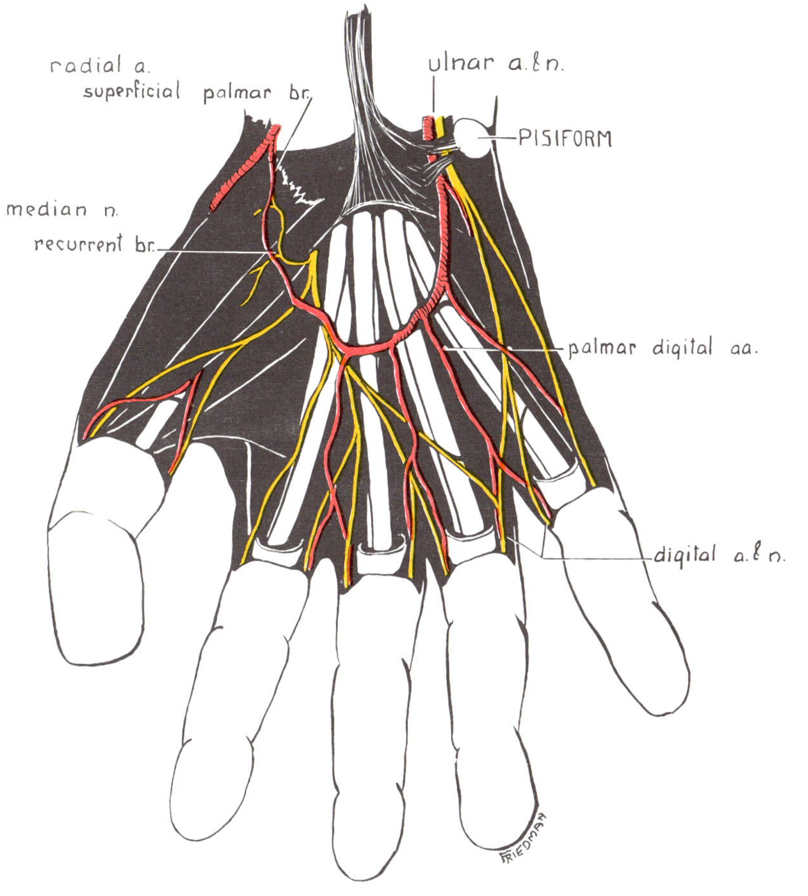
radial a.
superficial palmar br.
ulnar a. & n.
PISIFORM
median n.
recurrent br.
palmar digital aa.
digital a. & n.
FRIEDMAN

The Palmar Aponeurosis

The center of the palm is covered by a dense, triangular aponeurotic sheet, the **palmar aponeurosis.** This attaches proximally to the flexor retinaculum and receives the bulk of the tendinous fibers of Palmaris Longus when this muscle is present. Distally, the aponeurosis separates into bands which run to the lateral aspects of the carpometacarpal joints of the fingers. Transverse fibers link the palmar aponeurosis to the capsules of these joints superficially. On its medial side the aponeurosis blends with the dense fascia surrounding the hypothenar eminence; on its lateral side it blends with the thenar eminence.

Palmaris Brevis is a thin quadrilateral sheet of muscle which arises from the medial side of the palmar aponeurosis and from the adjacent part of the flexor retinaculum. The fibers pass over the front of the hypothenar eminence and insert into the overlying skin. It is innervated by the *ulnar nerve*. Its contraction wrinkles the skin of the hypothenar eminence and is said to tighten or brace the palmar aponeurosis and so help the grip.

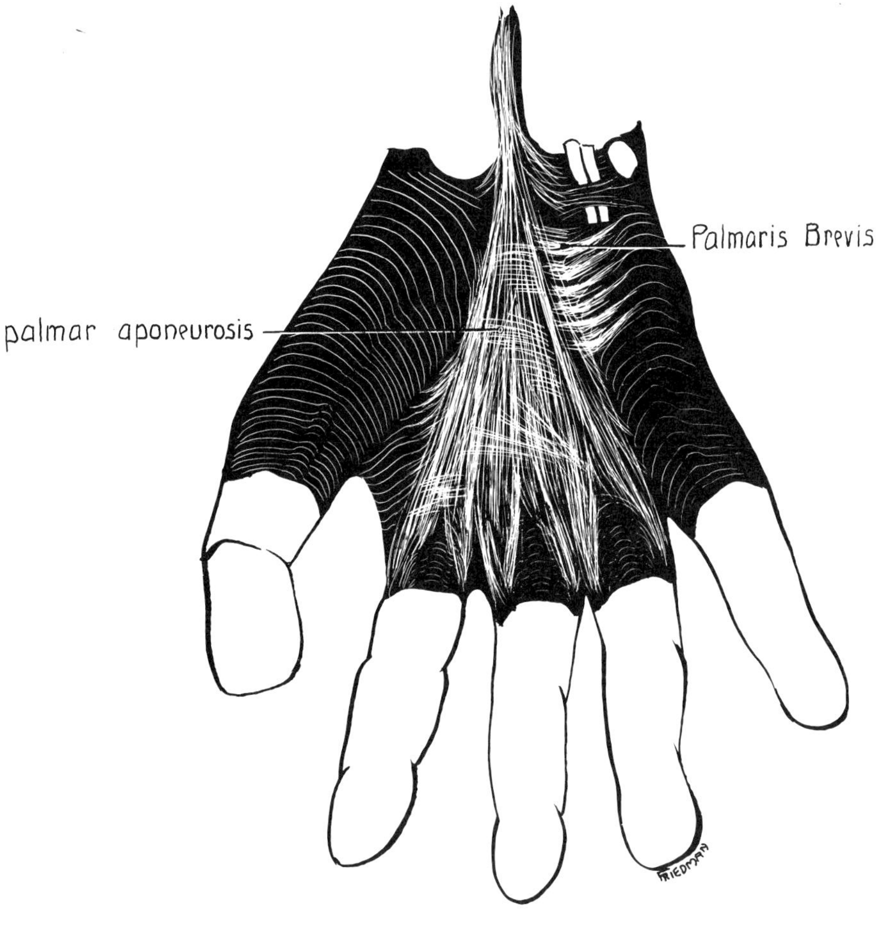
Palmaris Brevis
palmar aponeurosis
FRIEDMAN

The Back of Forearm and Hand: I. The Radius and Ulna

From a functional standpoint the back of the forearm begins at the posterior subcutaneous border of the ulna. The flexor compartment of the forearm extends around the medial side of the limb to this line and is represented here by Flexor Carpi Ulnaris. The flat, roughly triangular area above the posterior border is also subcutaneous, with a bursa intervening between bone and skin. Above this, the top of the olecranon receives the Triceps muscle. The remainder of the ulna is available for the extensor muscles.

It is essential to verify on oneself that the whole of the posterior border of the ulna is immediately subcutaneous and can be palpated throughout its length. This border demarcates the flexor compartment, which lies on its medial side, from the extensor compartment, which lies lateral to it.

Only the upper third of the posterior aspect of the radius is occupied, either by components of the joint capsule or by the Supinator muscle. All of the bone below this, together with the posterior aspect of the interosseous membrane, is available for extensors.

The deep branch of the radial nerve, after running through the Supinator muscle, enters this region as the **posterior interosseous nerve.** It innervates all of the extensor muscles except three of the superficial group arising from the common extensor origin and above; these three, Brachioradialis, Extensor Carpi Radialis Longus and Brevis, are innervated by the radial nerve before it enters Supinator. The **posterior interosseous artery,** a branch of the common interosseous of the ulnar, enters the region from above the interosseous membrane and concerns itself with the blood supply of the extensor muscles.

The lower end of the radius is marked by a series of grooves in which the tendons pass from forearm to hand.

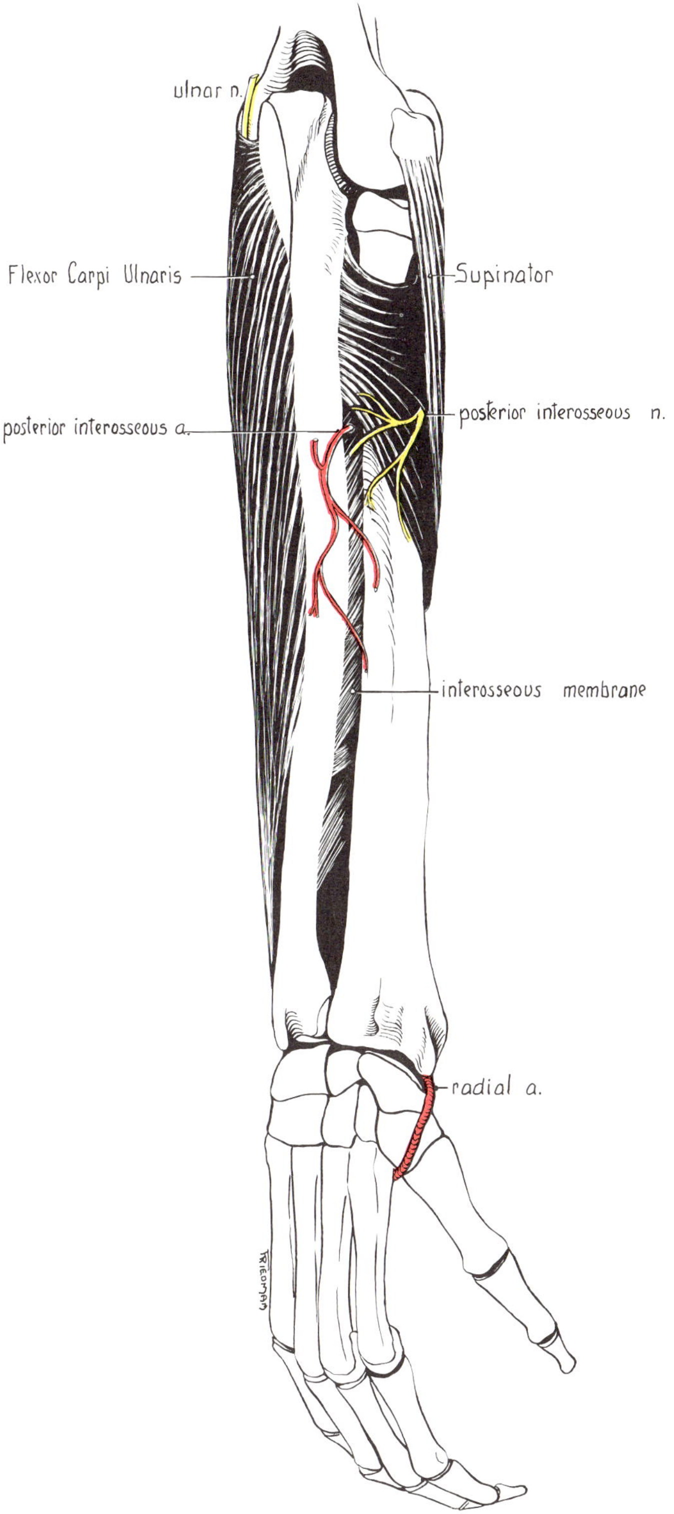
ulnar n.
Flexor Carpi Ulnaris
Supinator
posterior interosseous a.
posterior interosseous n.
interosseous membrane
radial a.

The Back of Forearm and Hand: II. The Muscles of the Extensor Compartment

Three muscles, innervated directly by the radial nerve, arise one above the other from the lateral epicondyle and the supracondylar ridge above it. Only the uppermost, the largest, is prominent in the fleshy part of the forearm. This is **Brachioradialis.** It arises from the upper half of the supracondylar ridge and forms a large, bulky, fleshy mass, which has already been seen as a prominent portion of the anterior aspect of the forearm (Fig. 24). There it forms the lateral margin of the cubital fossa and overlaps the radial artery for almost two thirds of the vessel's length. Inferiorly, it forms a small oval tendon which inserts into the lateral aspect of the styloid process of the radius. It thus acts only between humerus and radius and hence primarily on movements of the elbow. It is innervated by the *radial nerve* within the cubital fossa.

Underlying Brachioradialis is **Extensor Carpi Radialis Longus** and then **Extensor Carpi Radialis Brevis;** these run together vertically down the forearm. Their tendons pass in a common compartment grooving the posterior aspect of the radius. They then cross the back of the carpus and insert into the posterior aspect of the base of the second and third metacarpals. They thus match Flexor Carpi Radialis which inserts into the anterior aspect of the same two bones.

The three most deeply placed muscles of the extensor compartment of the forearm are all involved in modulating the complex movements of the thumb. The longest tendon, that of **Extensor Pollicis Longus,** passes all the way to the back of the distal phalanx and requires a pulley on the back of the radius to produce the correct alignment for extension. The pulley is readily felt as a *dorsal tubercle (Lister's)* on the back of the radius; the tendon beyond this point can be made to stand out by simple hyperextension of the thumb. The muscle fibers arise from the ulna.

The second of these long thumb muscles, **Extensor Pollicis Brevis,** inserts into the base of the first phalanx and requires no correction for its alignment. It runs to its destination from a small origin on the back of the radius. Running alongside it to insert into the lateral aspect of the base of the metacarpal is the third long thumb muscle, **Abductor Pollicis Longus,** which arises from both radius and ulna above its two companions.

Each long bone of the thumb thus has its own tendon inserting into its base. Since the tendon of Extensor Pollicis Longus is separated from the other two by a tubercle, there is an interval between it and the other two as they run across the carpus. This interval is recognizable as an indentation in the skin, the "anatomical snuff box," which is bounded anteriorly by the tendons of Extensor Pollicis Brevis and Abductor Pollicis Longus running together and posteriorly by that of Extensor Pollicis Longus running alone. These thumb tendons cross over those of Extensor Carpi Radialis Longus and Brevis. In so doing, these thumb muscles pass from the deepest to the most superficial plane, a fact that has led to their being named "outcropping thumb muscles" (Grant).

The small **Extensor Indicis** muscle which also arises from the deep plane will be described in Figure 41.

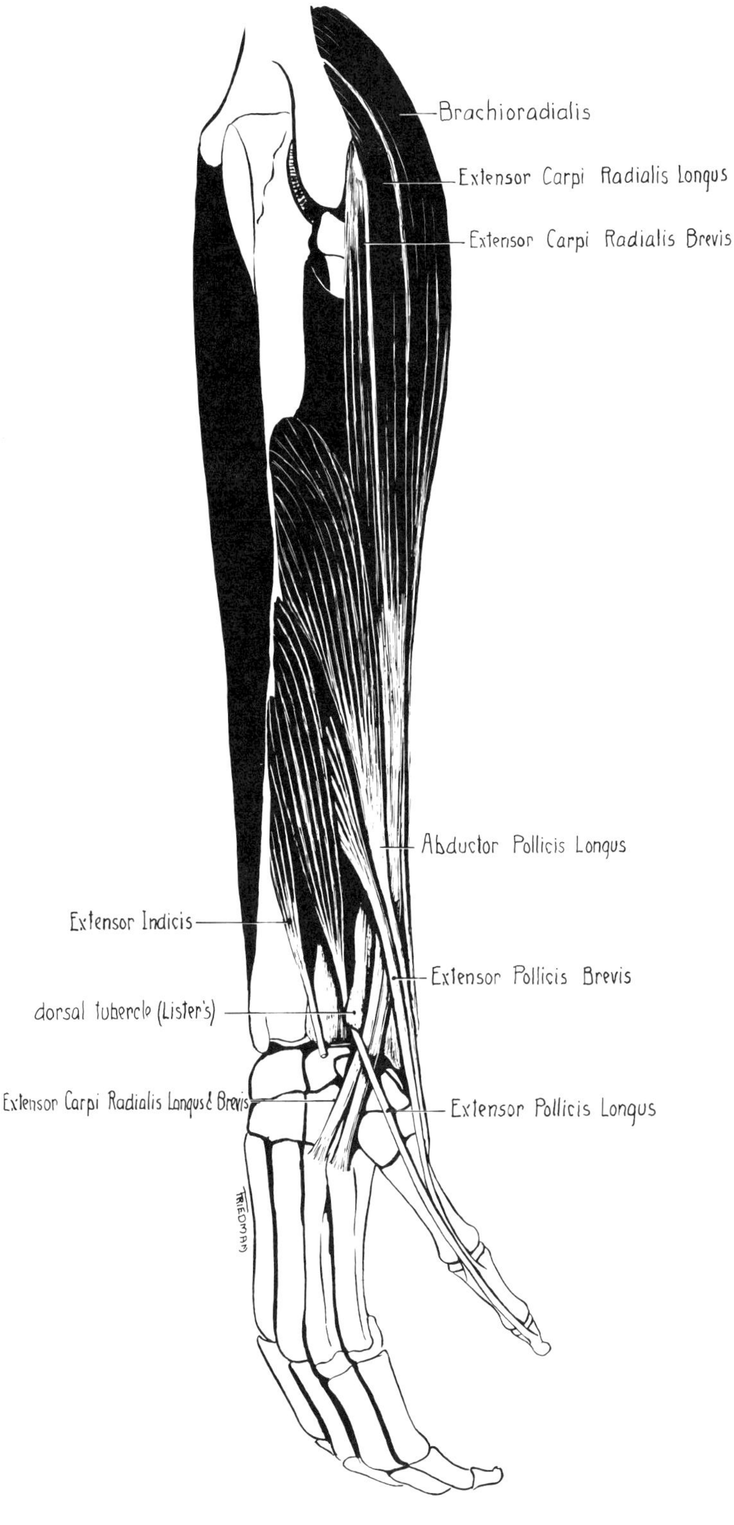
Brachioradialis
Extensor Carpi Radialis Longus
Extensor Carpi Radialis Brevis
Abductor Pollicis Longus
Extensor Indicis
Extensor Pollicis Brevis
dorsal tubercle (Lister's)
Extensor Carpi Radialis Longus & Brevis
Extensor Pollicis Longus

The Back of Forearm and Hand:
III. The Muscles of the Extensor Compartment (cont'd)

The superficial muscle group begins with Brachioradialis and includes Extensor Carpi Radialis Longus and Brevis even though these are largely overlapped by Brachioradialis. The area of origin for the superficial extensors thus begins with the lateral supracondylar ridge and continues down to include the lateral epicondyle. From here, the extensor origin passes to the back of the humeral condyle—specifically, the area behind the capitulum where a group of fleshy extensors takes origin in common.

The most superior member of the series is **Anconeus.** This triangular muscle inserts into the lateral aspect of the upper quarter or so of the ulna. Below this is **Extensor Carpi Ulnaris,** the counterpart of Flexor Carpi Ulnaris. Like the latter, it arises from a shared origin on the humerus as well as from the posterior subcutaneous border of the ulna. It passes down and crosses the wrist in a groove between the head and styloid process of the ulna to insert into the posterior aspect of the base of the fifth metacarpal—that is, opposite to the insertion of Flexor Carpi Ulnaris.

Extensor Digitorum Communis lies alongside Extensor Carpi Ulnaris, between it and the Extensor Carpi Radialis group. The muscle splits into tendons which produce extension of the fingers through the extensor expansion as will be described. There are tendons to each of the fingers. A separate additional muscle belly and tendon for the little finger are frequently observed and named **Extensor Digiti Minimi.** The extensor tendon to the index finger is joined on its medial aspect by the **Extensor Indicis,** which, arising with the deep group of thumb muscles from the lower end of the radius, surfaces to meet the tendon below the wrist.

At this point it is convenient to summarize some of the features of the muscles of the wrist proper, i.e., those with "carpi" in their name. The insertion of these muscles is similar on both front and back of the wrist, i.e., into the bases of the second, third, and fifth metacarpals front or back. The flexor group arises from the common flexor origin, the extensor group from the common extensor origin. Both Flexor and Extensor Carpi Ulnaris extend their origins to include the posterior border of the ulna as well. Flexor Carpi Radialis divides into two tendons only at its insertion but evidently corresponds to Extensor Carpi Radialis Longus and Brevis, which are separate throughout. Acting together, the anterior group flexes the wrist, the posterior group extends the wrist. The flexors and extensors may also act in concert on either the ulnar or radial side to produce adduction or abduction and may act serially to rotate the wrist.

The **extensor retinaculum** is a fascial retaining band around the back of the wrist which extends from the lower end of the radius to the lower end of the ulna and adjacent carpal bones. Synovial-lined compartments of the retinaculum associated with grooves on the back of the radius and ulna restrain the tendons in groups.

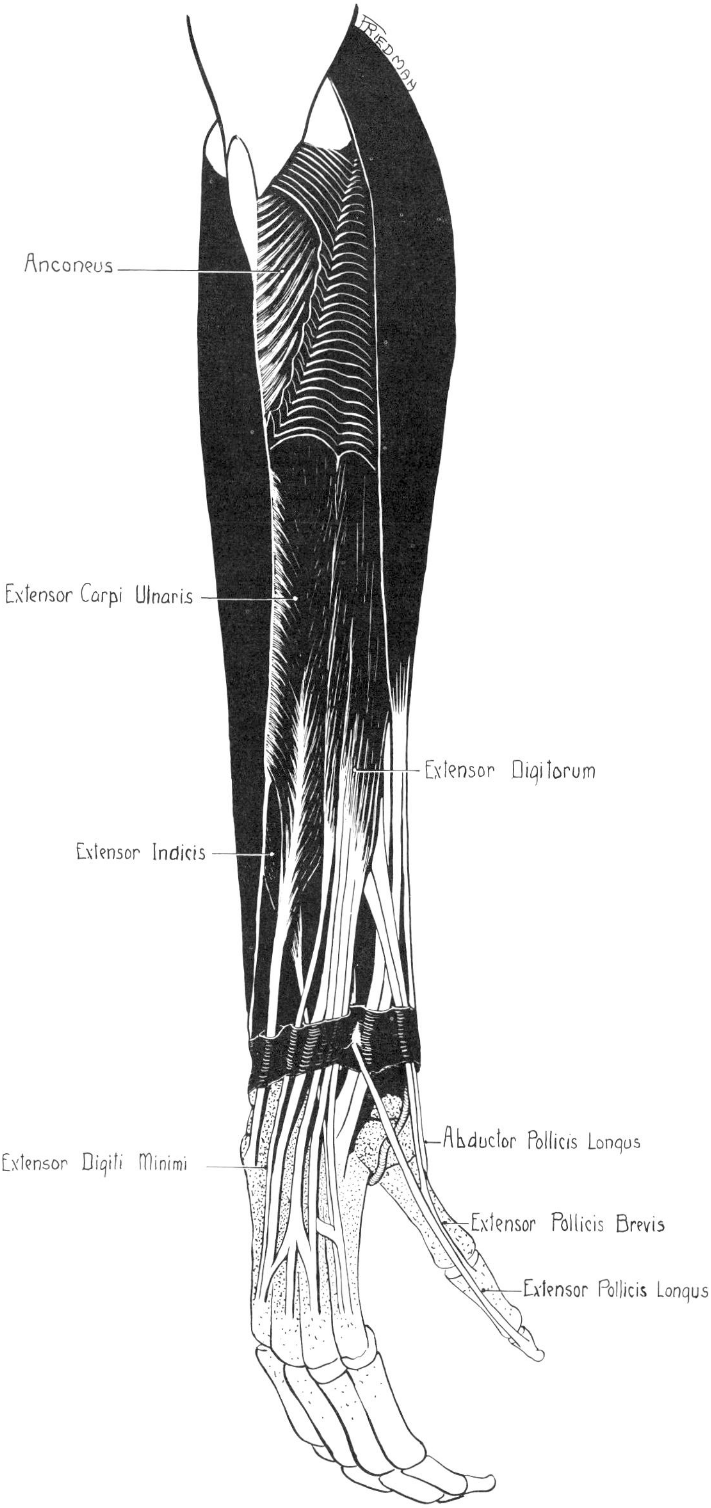

FRIEDMAN
Anconeus
Extensor Carpi Ulnaris
Extensor Digitorum
Extensor Indicis
Abductor Pollicis Longus
Extensor Digiti Minimi
Extensor Pollicis Brevis
Extensor Pollicis Longus

The Back of Forearm and Hand: IV. The Extensor Expansion

The **metacarpophalangeal** and the **interphalangeal** joints are each surrounded by a simple fibrous capsule which attaches a short distance beyond the articular margins in each case. Within this fibrous capsule is a simple synovial membrane ensheathment. The *collateral ligaments* of the metacarpophalangeal joints merit special mention. These bands pass from the nonarticular portion of the head of the metacarpal into the base of the phalanx beyond. They are eccentrically placed so that they become taut in flexion and loose in extension of the digits. This can easily be verified in the living subject; in the extended position, the digit can be moved quite freely at the metacarpophalangeal joint, whereas scarcely any side-to-side movement is possible when this joint is flexed.

The *deep transverse metacarpal ligament* joins adjacent anterior portions of the metacarpophalangeal joint capsules. It provides a fulcrum for the Lumbrical muscles which, passing in front of it, continue into the dorsum of the fingers. Acting at this fulcrum, the first movement of a Lumbrical is to aid in flexion at the metacarpophalangeal joint.

All of the tendons that pass to the back of a finger insert in one way or another through a common fibrous hood which covers the dorsal aspect of the first phalanx and extends by slips into the middle and distal phalanges. This **extensor expansion** or **dorsal aponeurosis** is attached by its middle fibers to the dorsal aspect of the proximal part of the second phalanx of each digit. On either side of this it is prolonged by lateral slips which, uniting, insert into the dorsal aspect of the distal phalanx. The Lumbrical muscle inserts into the lateral aspect of the basal portion of this expansion. Acting at this point, it first flexes the digit and then, the pull of its fibers prolonged through the lateral portion of the expansion, produces extension at the interphalangeal joints.

In similar fashion, the Interosseus muscle, be it Dorsal or Palmar, passes posterior to the deep transverse metacarpal ligament to insert in part directly into the bone of the proximal phalanx and in part into the inferolateral portion of the expansion. These muscles thus produce not only side-to-side movement of the digits at the metacarpophalangeal joint but, acting distally through the expansion, aid in extension at the interphalangeal joints.

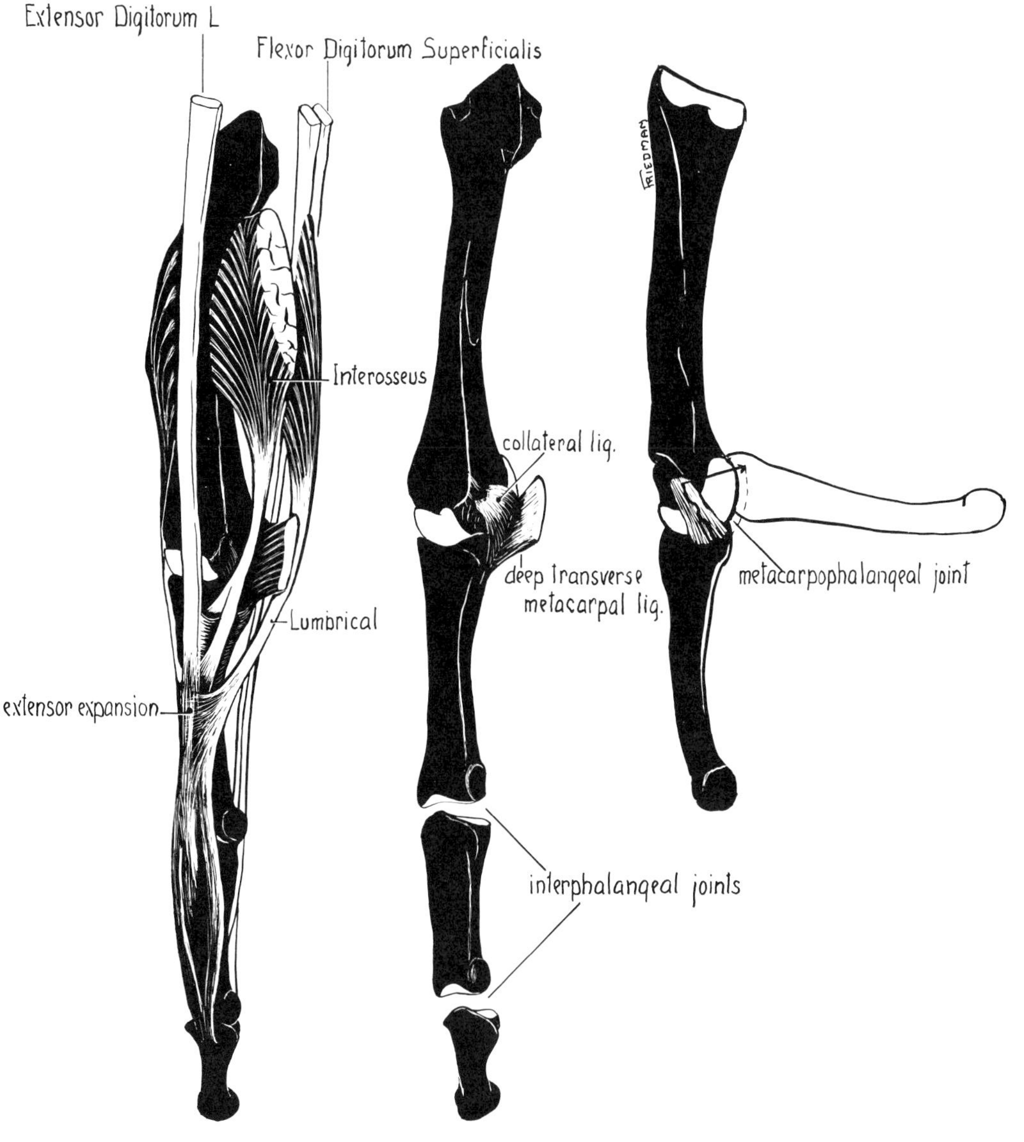
Extensor Digitorum L
Flexor Digitorum Superficialis
Interosseus
Lumbrical
extensor expansion
collateral lig.
deep transverse metacarpal lig.
metacarpophalangeal joint
interphalangeal joints
FRIEDMAN

Surface Features of the Upper Limb

The Mammary Gland

Whether it is described as part of the thorax or in connection with the pectoral portion of the upper extremity, the most obvious surface feature of the region is the **mammary gland.** This is a skin gland enlarged and modified to serve the specific function of lactation. In the male it normally remains rudimentary throughout life. It consists of 15 to 20 individual lobes on each side. Each lobe is a cluster of secreting alveoli united by ducts which progressively increase in size until they form a final single channel, the *lactiferous duct.* Each of these 15 to 20 ducts conveys its secretion to the summit of the skin mound, the *nipple.* A small enlargement of the duct close to its termination is the *lactiferous sinus.* Each of these lobes or clusters is separated from its fellow by a more or less complete fibrous septum so that the lobes are arranged radially like the spokes of a wheel.

Each lobe is protected by an extensive packing of adipose tissue retained in position by numerous fascial septa which are anchored to the overlying skin, the *suspensory ligaments (of Cooper).* Surrounding the nipple is a pigmented disc of skin, the *areola,* pink in the virgin but becoming large and purplish in pregnancy. Small glands open onto the surface of the areola and provide an oily secretion which serves to lubricate the nipple during lactation. The nipple is vascular and contains a considerable amount of smooth muscle, features that combine to make it erectile.

Since there are no rigid restraints, it is difficult to define the extent of the "normal" mammary gland. In youth it extends roughly between the second and sixth ribs and from the sternum into the axilla. Unsupported, the breast tends to become increasingly pendulous with age. The actual glandular material in the nonpregnant state is considerably less in amount than surface appearance of the fat-packed mass suggests. It is of importance that in addition to the approximately circular outline of the gland a *tail* extends upward around the lower margin of Pectoralis Minor into the axilla.

The mammary gland receives its supply of raw material largely through the *lateral thoracic artery,* a branch of the second part of the axillary. This artery becomes enormously enlarged during pregnancy. It receives additional blood from the *intercostal arteries* supplying the adjacent intercostal spaces.

The lymphatic drainage is of prime importance, since it is along these pathways that malignant tumors of the breast extend. The gland is encased on all its surfaces by a net of lymphatic channels. These channels coalesce along its deep surface—i.e., between the gland and Pectoralis Major—and superficially around the areola. From here, major lymphatic channels radiate peripherally, as will be described in the next illustration.

The mammary gland develops from ectoderm along the *milk ridge,* which extends obliquely from the axilla toward the groin. Supernumerary glands, *polymastia,* or nipples, *polythelia,* may occasionally develop along this line.

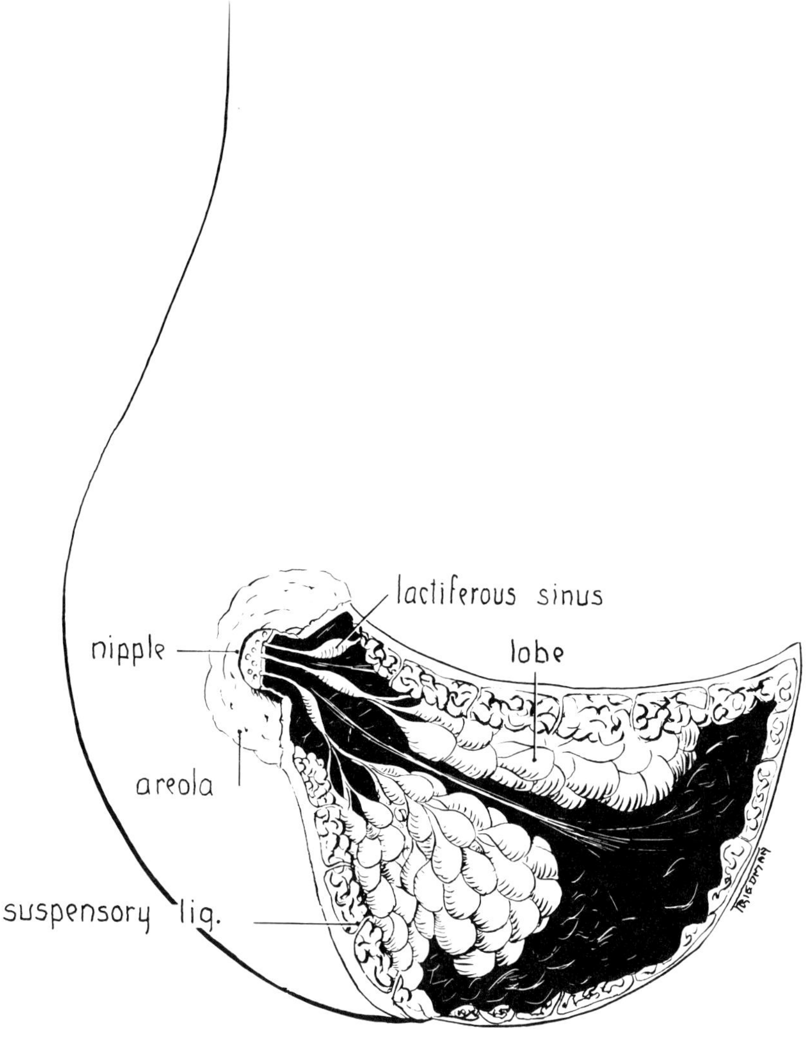
lactiferous sinus
nipple
lobe
areola
suspensory lig.

The Superficial Veins and Lymphatics

The large veins underlying the skin of the upper extremity are very accessible to needle and catheter. The superficial parts of both the front and the back of the hand drain into major channels usually located singly, one on either side of the anterior aspect of the forearm. Laterally, the *cephalic vein* runs along the lateral margin of the cubital fossa and the lateral border of Biceps to the interval between Deltoid and Pectoralis Major. Here it is conducted to the deltopectoral triangle, where it pierces the underlying clavipectoral fascia and usually ends in the subclavian vein. Medially, the *basilic vein* passes up along the medial side of the cubital fossa, pierces the deep fascia above the elbow, and joins the brachial veins. Cephalic and basilic veins are cross-connected in front of the fascia that overlies the cubital fossa. The cross channel usually receives a *median vein,* which brings blood from the deeper portion of the forearm. The cephalic and basilic veins and their cross connections are particularly accessible at the elbow. The veins on the dorsum of the hand are also easily entered.

For the most part, the lymphatic drainage of the upper limb follows the pattern of the veins. The lymphatics from the front of the hand pass up the forearm in vertical channels. They receive drainage from the dorsum of the hand near the wrist and a continual accretion of channels from the dorsum of the forearm as they pass upward to approach the medial side of the Biceps tendon. Here there is usually an important lymph node, the *supratrochlear,* at the elbow in front of the medial epicondyle.

From the elbow, channels pass with the brachial veins up the medial side of the arm and continue along through the axilla both inside and outside the axillary sheath and ultimately enter into a confluence of lymphatics at or near the junction of the internal jugular vein with the subclavian vein. On the right, the final channel is usually the **right lymphatic duct;** on the left, the **thoracic duct.**

Collections of lymph nodes interrupt the lymphatic pathways within the axilla; these are identified by clinicians in relation to the basic structures with which they are associated. The principal named groups are thus *axillary,* those surrounding the axillary artery and vein; *subscapular,* along the course of the subscapular vessels; *pectoral,* along the lower border of Pectoralis Minor and adjacent chest wall in association with the lateral thoracic artery; and finally, a somewhat loosely identified *central* group, located more or less in the center of the axilla. For those lymphatics that follow the cephalic vein there are also *infraclavicular* nodes located in the deltopectoral triangle.

The lymphatics draining most of the mammary gland pass from the periphery along its tail into the pectoral group of lymph nodes and proceed from there through the central group into the terminal lymphatic channels. In addition, the upper part of the gland drains superiorly into the infraclavicular group. The medial margin drains into the channels and nodes alongside the internal thoracic artery. From the inferomedial quadrant some channels also pass down toward the rectus sheath; and most importantly, a few pass directly into the mediastinum.

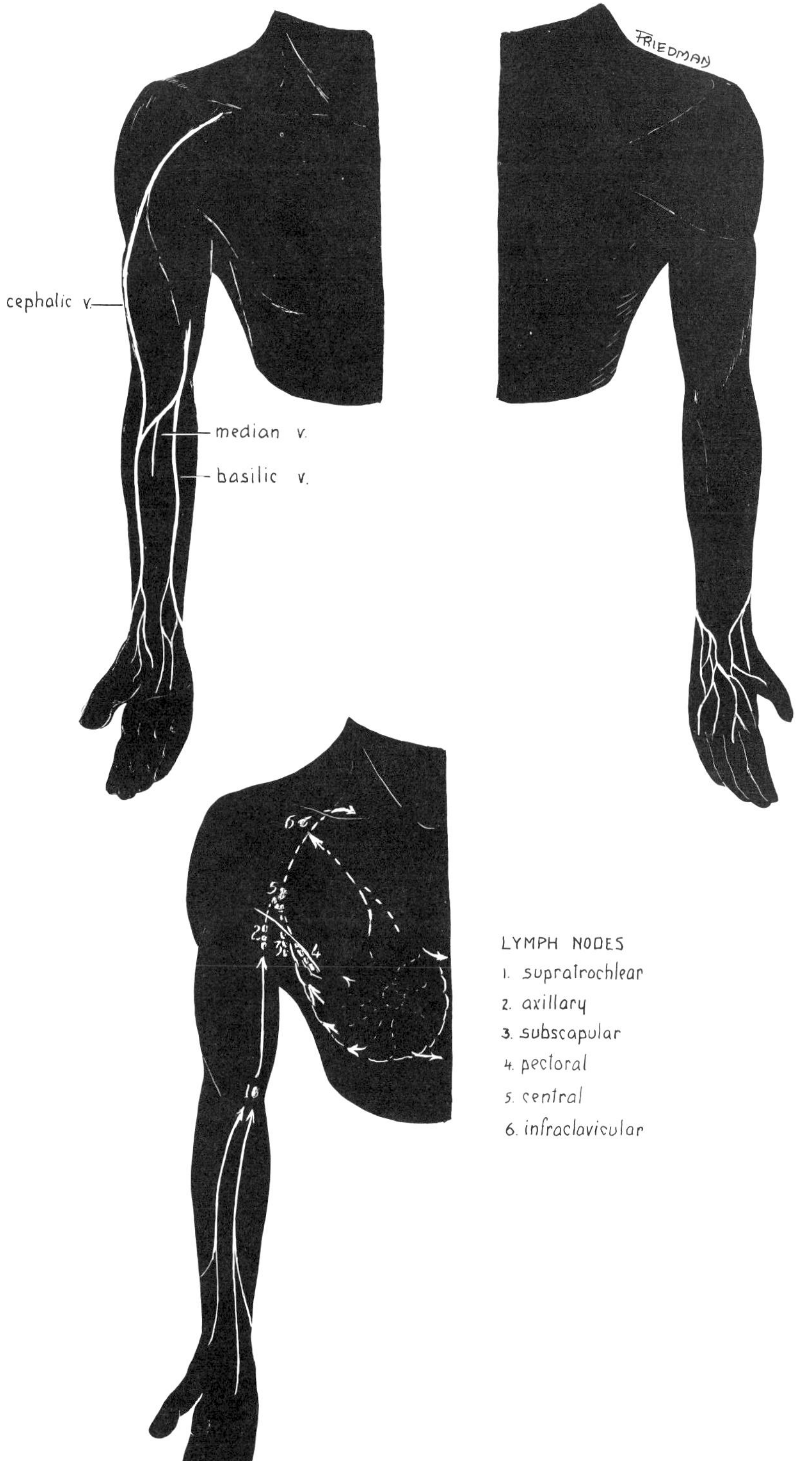
FRIEDMAN
cephalic v.
median v.
basilic v.
LYMPH NODES
1. supratrochlear
2. axillary
3. subscapular
4. pectoral
5. central
6. infraclavicular

The Dermatomes

The important, detailed distribution of motor nerves to the muscles of the upper limb has been described in connection with individual muscle groups. From the clinical standpoint, some features of the cutaneous innervation of the skin overlying the upper extremity may on occasion provide clues to the location of a nerve lesion or may be important in indicating the appropriate procedure for securing a regional anesthesia. These cutaneous areas, or dermatomes, overlap one another extensively, and only the broadest outline is here attempted.

Useful generalizations can be made about spinal cord segments. Thus, the region of the shoulder is innervated by C3, 4; the lateral aspect of the arm and forearm by C5, 6; the lateral half of the hand roughly by C6, 7. The medial side of the hand is served by C7, 8; and the medial side of the forearm and arm by C8 and T1. This pattern of distribution depends upon the fact that, embryologically speaking, the upper extremity develops as if the limb, thumb up, were extended straight out laterally in the horizontal position, pushing its encasing skin before it.

In specific terms, the region of the cap of the shoulder is supplied by the supraclavicular nerves which reach it from the posterior triangle of the neck. The area, both front and back, around the middle of the Deltoid muscle is innervated by the axillary nerve, while the skin around the lower part of the muscle receives fibers from the radial nerve. The musculocutaneous nerve looks after the lateral side of the lower arm and forearm, while the corresponding portions on the medial side of arm and forearm are supplied by the medial cutaneous nerve of the arm and the medial cutaneous nerve of the forearm respectively, both of which arise from the medial cord of the brachial plexus in the axilla. The medial side of the arm is also innervated by the *intercostobrachial nerve* from the 2nd and 3rd intercostal nerves (*Visual Anatomy II: Thorax and Abdomen, Fig. 6*).

The anterior aspect of the hand receives innervation for its lateral two thirds from the median nerve and for its medial one third from the ulnar. Both of these nerves extend to the back to innervate the skin of the dorsum of the distal phalanges. The dorsum of the hand receives innervation for its lateral two thirds from the radial nerve and for its medial one third from the ulnar.

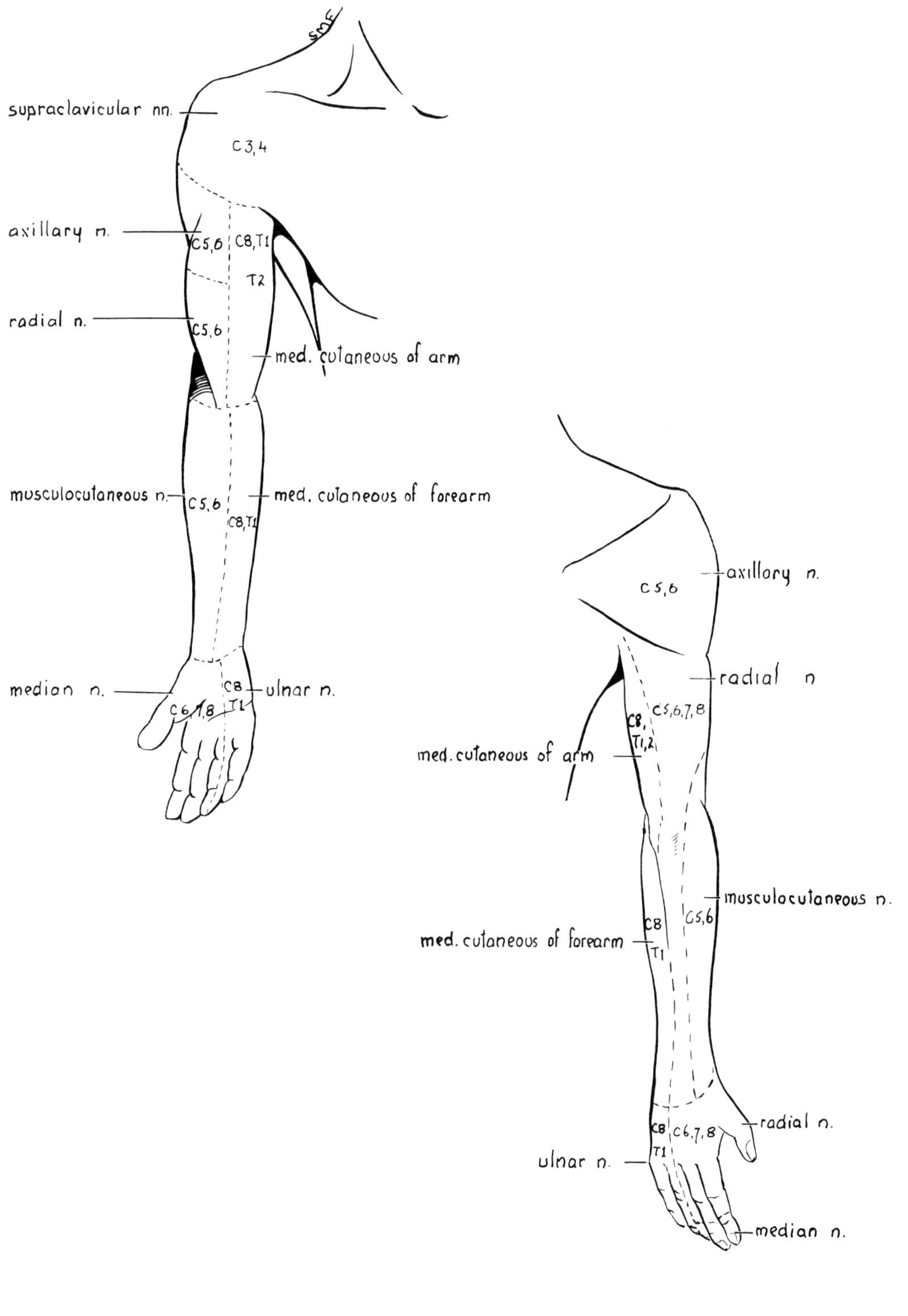
supraclavicular nn.
C3,4
axillary n.
C5,6
C8,T1
T2
radial n.
C5,6
med. cutaneous of arm
musculocutaneous n.
C5,6
C8,T1
med. cutaneous of forearm
median n.
C6,7,8
C8
T1
ulnar n.
axillary n.
C5,6
radial n
C5,6,7,8
C8,
T1,2
med. cutaneous of arm
musculocutaneous n.
C5,6
C8
T1
med. cutaneous of forearm
radial n.
C8
C6,7,8
T1
ulnar n.
median n.

The Suspension of the Lower Limb

The Framework:
I. The Anterior Aspect of the Hip Bone

In this illustration, the **hip bone** is seen obliquely from the front, rotated sufficiently to bring the acetabulum, which receives the head of the femur, into view. In the anatomical position the acetabulum is directed laterally.

The hip bone is formed of ilium, ischium, and pubis—three independent bones which meet one another in the acetabulum, each forming about a third of it. They begin to unite during adolescence and fuse completely in early adulthood (20 to 25 years). The **ilium** extends upward as a flat, shallowly dished bone, placed almost at right angles to the pubis and ischium. This right-angled bend is greatly reinforced along the inner aspect of the hip bone, and it is this reinforcing mass that forms the *brim of the pelvis.* The *iliac crest* surmounts the bone and provides origin for the muscles of the anterior abdominal wall. It terminates anteriorly in the *anterior superior iliac spine,* a basic landmark in physical examination. A second small knob, the *anterior inferior iliac spine,* provides an origin for Rectus Femoris a short distance below the superior spine.

The portions of **ischium** and **pubis** below and in front of the acetabulum provide areas for muscle attachments, as will be described, but are less massive than the ilium, since they are not directly involved in transferring the weight of the trunk through the hip bone to the femur. The rounded brim of the pelvis sharpens anteriorly as the *pectineal line of the pubis* and ends at the *pubic tubercle,* another basic surface landmark. The large bony deficiency between pubis and ischium, the *obturator foramen,* is further evidence that these portions of the hip bone are not needed for weight bearing.

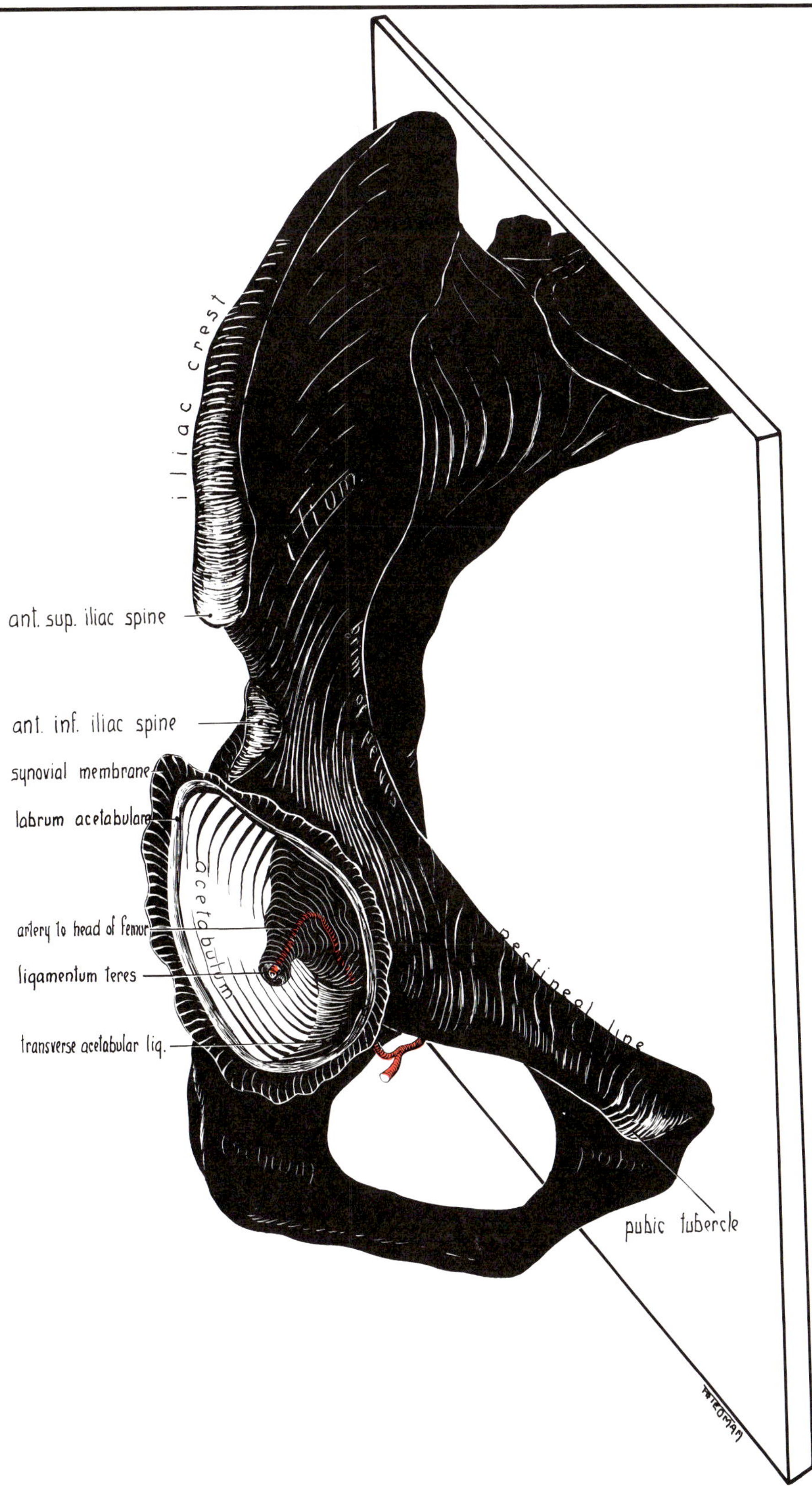
iliac crest
ilium
ant. sup. iliac spine
brim of pelvis
ant. inf. iliac spine
synovial membrane
labrum acetabulare
acetabulum
artery to head of femur
ligamentum teres
pectineal line
transverse acetabular lig.
ischium
pubis
pubic tubercle

The Framework:
II. The Posterior Aspect of the Hip Bone

The angulation of the upper half of the hip bone with respect to the lower half is very apparent when the bone is viewed from behind. So too, the manner in which the weight of the body is transferred through the ilium across the acetabulum into the femur is readily appreciated.

The outer or gluteal surface of the iliac bone is almost wholly concerned with providing attachments for the gluteal muscles, which are separated from one another by the *inferior, anterior,* and *posterior gluteal lines*. The posterior part of the ischium, the *ischial tuberosity,* together with the area immediately below it, provides attachment for the hamstring muscles. Between these areas of muscle attachment is the *ischial spine,* which demarcates the *greater sciatic notch* above from the *lesser sciatic notch* below. These notches are the major access routes between the pelvis and the back of the thigh and buttock.

Muscles that run between the gluteal surface of the hip bone and the greater trochanter of the femur are placed so as to approximate the two. Thus, for example, if one foot is on the ground as in walking or skating, the whole body can be tilted up, over, and forward. This movement is conventionally described as abduction of the lower limb, as indeed it is when the trunk is fixed in space. It is quite clear, however, that movement of the trunk on the limb fixed on the ground occurs at least as frequently. To the extent that muscles responsible for such movement pass either behind or in front of the axis of the joint, the trunk (or lower limb) will be rotated as well. These are, then, the typical actions of the gluteal muscles, readily displayed by swiveling the body while standing on one limb.

In a similar but opposite sense, muscles that arise from the outer aspect of the lower elements of the hip bone and run to the shaft of the femur adduct the lower limb. This movement, in contrast to abduction, is aided by gravity; this is reflected in the smaller size of these muscles.

The movements of the hip joint are generally described in terms of abduction and adduction, flexion (forward), and extension (backward). Combined together, the movement produced is circumduction. There is an enormous variability in the freedom of these movements from one individual to another. Although they are all described in terms of the lower limb, they are used at least as often to move the trunk on the limb, as in walking. This should be kept constantly in mind while studying this region.

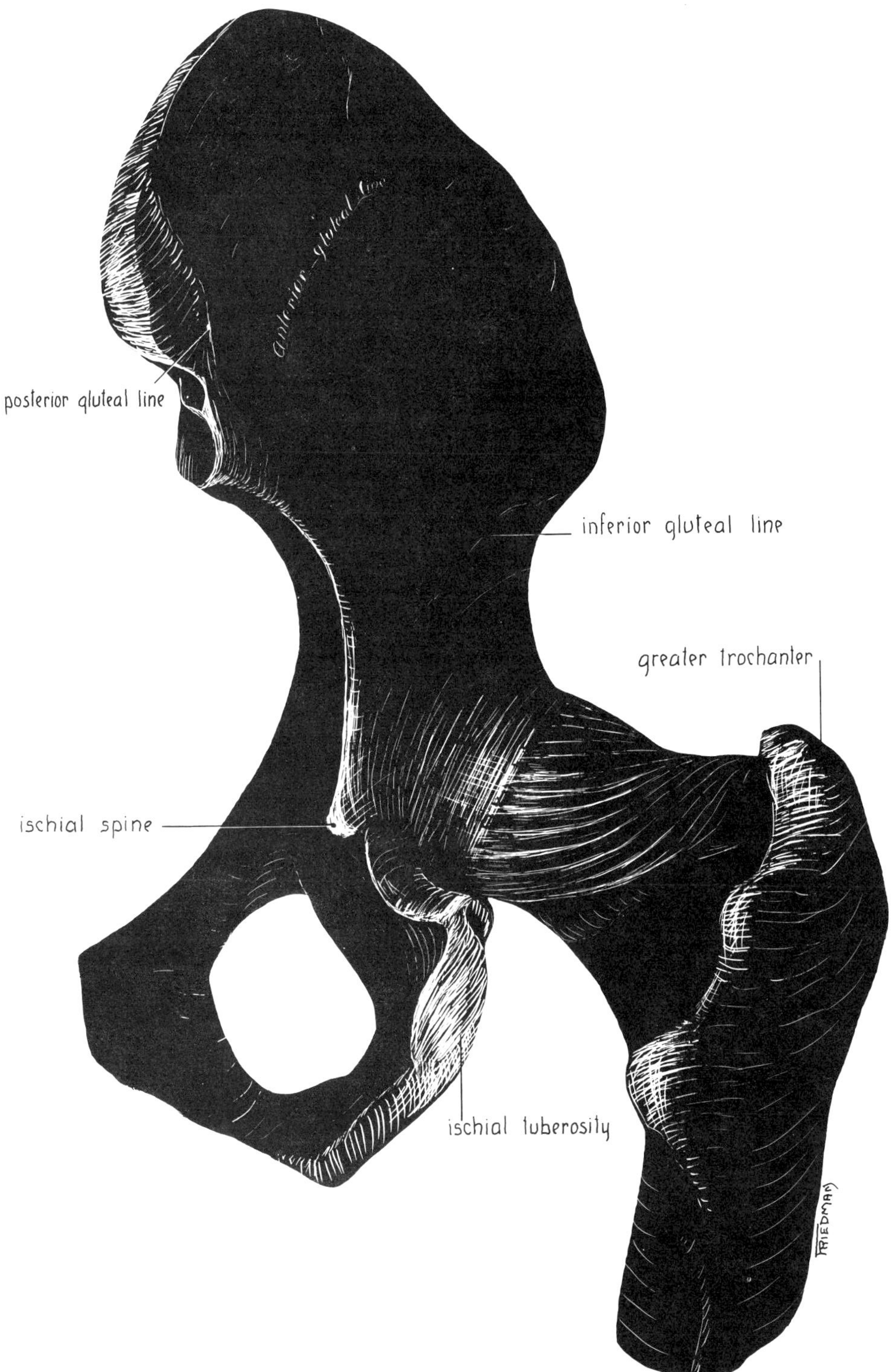
anterior gluteal line
posterior gluteal line
inferior gluteal line
greater trochanter
ischial spine
ischial tuberosity
FRIEDMAN

The Hip Joint:
I. The Synovial Membrane

In contrast to the upper end of the humerus, the head of the femur is separated from its shaft by a neck of sizable proportions. As in the humerus, however, the upper end of the shaft of the femur expands to form large muscle-anchoring protuberances, the greater and lesser trochanters. The *greater trochanter* is a large, laterally placed mass of bone at the upper end of the femur and lies almost directly under the skin. The *lesser trochanter* is placed medially below and behind the neck, and cannot ordinarily be felt. Anteriorly, a rough *intertrochanteric line* joins the two trochanters. Posteriorly, they are joined by a prominent crest of bone, the *intertrochanteric crest.* The head of the femur is slightly more than a half sphere and is indented anteroinferiorly by a pit that receives the artery of the head together with its associated ligament and synovium.

The **synovial membrane** attaches to the articular margins of the head and continues laterally to cover the neck. It is then reflected off at the intertrochanteric line anteriorly; but posteriorly, since the neck is longer than it is in front, the membrane fails to meet the intertrochanteric crest before reflection. Thus, a fracture through the neck may tear through the synovial membrane and into the joint anteriorly but may pass wholly outside the synovial joint posteriorly.

The main arterial blood supply to the head and neck of the femur reaches the neck both from below and from behind and originates from the adjacent arteries such as the medial and lateral circumflex femoral. Branches pass linearly between bone and synovium, and form visible ridges, *retinacula,* in the latter. These are most apparent on the superior part of the neck and are called "retinacular arteries" by surgeons.

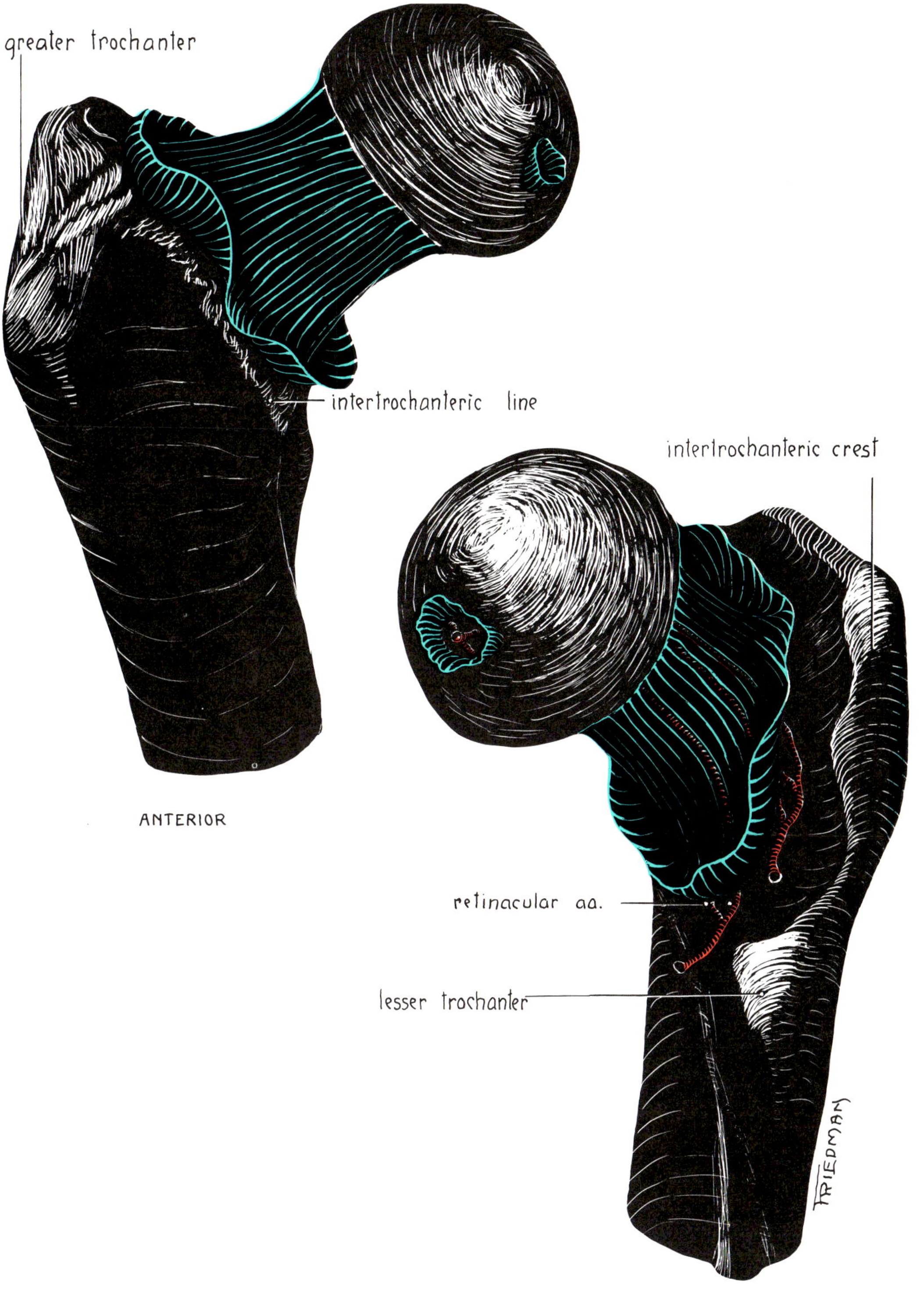
greater trochanter
intertrochanteric line
ANTERIOR
intertrochanteric crest
retinacular aa.
lesser trochanter
FRIEDMAN
POSTERIOR

The Hip Joint:
II. The Fibrous Capsule

The major part of the fibrous capsule of the hip joint surrounds the neck of the femur, extending from the acetabulum to the intertrochanteric line anterolaterally. It does not attach to the femur either posteriorly or inferiorly, but spirals about these parts of the neck with the result that it becomes tighter as the femur is extended. Thus, the anterior portion of the capsule, the *iliofemoral ligament,* runs from the ilium downward and laterally across the front of the joint to attach to the intertrochanteric line of the femur. Below this, the *pubofemoral ligament* arises from the pubis directly in front of the acetabulum and spirals below the joint to run into the back of the capsule itself. Here it is joined by the *ischiofemoral ligament,* which arises behind the acetabulum and spirals up and forward over the top of the joint to blend with the anterior portion of the capsule. An orbicular band of fibers, the *zona orbicularis,* marks out the free lower and posterior margins of the joint capsule.

It is apparent that if the pelvis is tilted backward at the hip joint—as, for example, in standing at ease—the capsular ligaments are tightened and tend to screw the femur more tightly into the acetabulum. This stance transfers the responsibility for stabilizing the bony skeleton away from muscle, with its attendant possibility of fatigue, to ligaments, which do not tire.

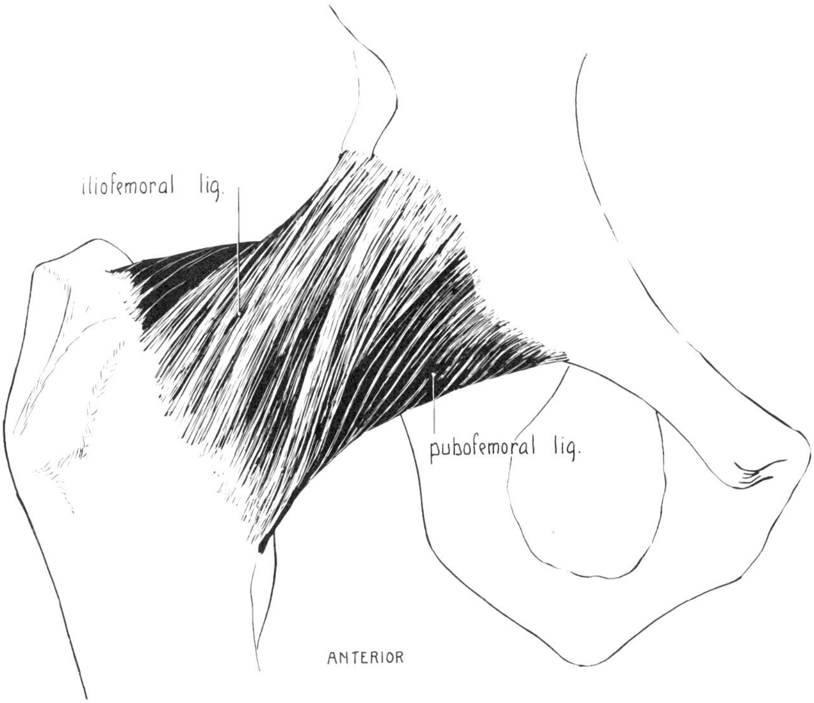
iliofemoral lig.
pubofemoral lig.
ANTERIOR

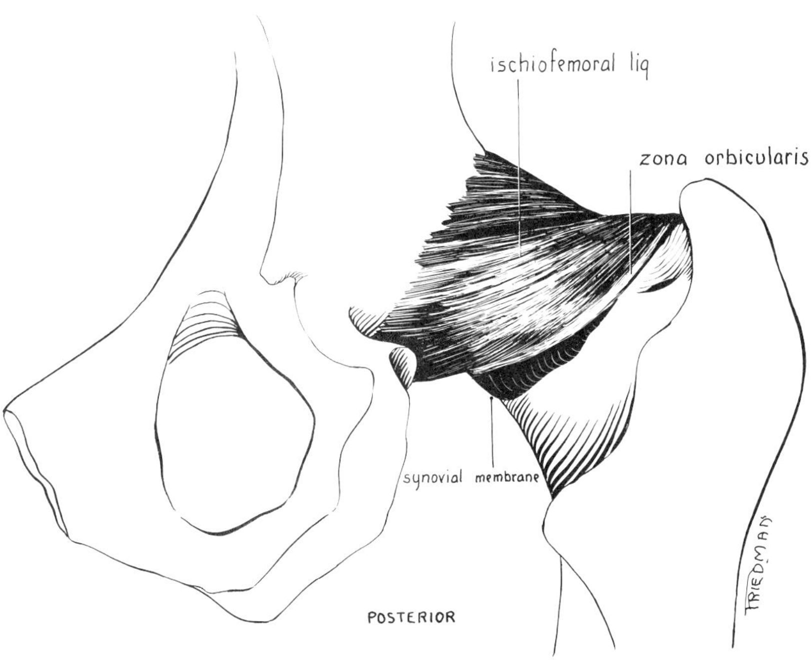
ischiofemoral lig
zona orbicularis
synovial membrane
POSTERIOR

Muscular Union of the Hip Bone and Femur

As in the case of the shoulder joint, a cuff of muscles closely applied to the capsule provides an elastic restraint for the hip joint.

Piriformis arises from the anterior surface of the sacrum adjacent to the upper three anterior sacral foramina and passes out of the pelvis through the greater sciatic foramen, which it almost entirely fills. The muscle tapers rapidly to a tendon which passes across the upper part of the joint capsule to insert into the highest surface of the greater trochanter.

Obturator Externus arises from the outer aspect of the obturator membrane as well as from its adjacent bony margin and spirals below the joint capsule, then up around and behind it to insert into a pit just medial to the intertrochanteric crest. **Obturator Internus** arises from the internal aspect of the obturator membrane and its adjacent bone, passes out of the pelvis through the lesser sciatic foramen, and here makes an almost right-angled bend to continue as a stout tendon across the back of the joint capsule. It inserts into the femur just above Obturator Externus. The tendon is reinforced and supported by the **Superior** and **Inferior Gemellus,** muscles that arise from the lesser sciatic notch just above and below the tendon of Obturator Internus and accompany it along its upper and lower borders to its destination.

The front of the capsule is reinforced by the conjoined tendon of **Iliacus** and **Psoas (Iliopsoas),** which passes obliquely down across the front of the joint to insert into the lesser trochanter of the femur and the area immediately below this. It will be described more fully in connection with the adductor group of muscles.

All of these muscles are closely applied to the joint capsule. Like the capsular ligaments, they too are all put under stretch when the joint is extended. A detailed description of Piriformis, Obturator Internus, Iliacus, and Psoas is given in *Visual Anatomy II: Thorax and Abdomen.*

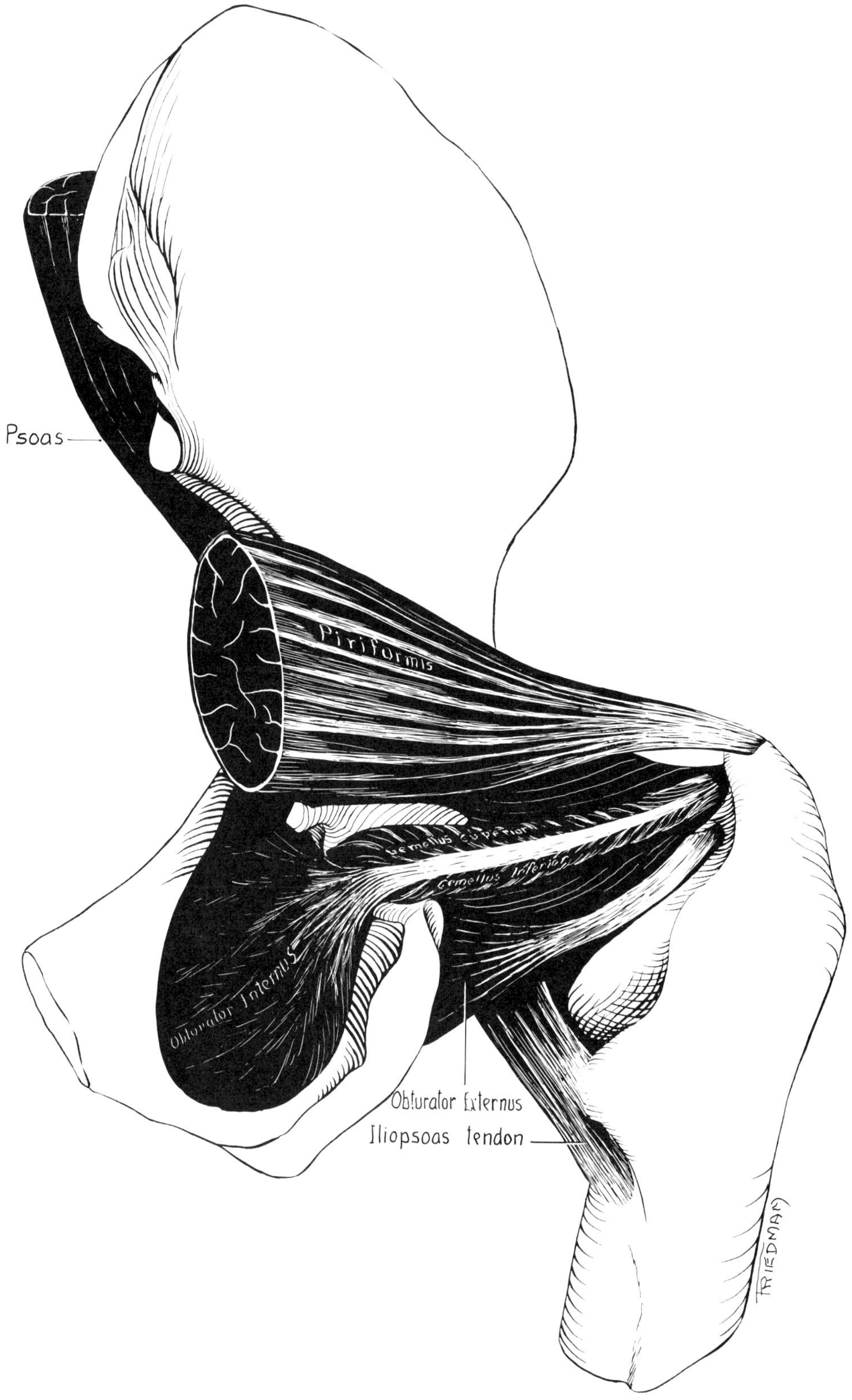
Psoas
Piriformis
Gemellus superior
Gemellus inferior
Obturator internus
Obturator externus
Iliopsoas tendon
FRIEDMAN

The Gluteal Region:
I. The Deep Plane of Muscle

The gluteal region is the area covered over by Gluteus Maximus, the large superficial muscle of the buttock. The lateral limit of the area is the greater trochanter of the femur; its medial limit is the ischial tuberosity, sacrotuberous ligament, and the midline of the sacrum.

Piriformis is the key muscle of the region. Its fleshy mass fills the greater sciatic foramen, but it quickly tapers to a tendon which passes to the high point on the greater trochanter of the femur. It is innervated by S2 and S3 fibers from within the pelvis. All of the nerves, arteries, and veins that leave the pelvis for the lower limb do so by passing, tightly packed, out of the greater sciatic foramen below Piriformis, except for the superior gluteal artery and nerve, which emerge above Piriformis. Not surprisingly a notable complication of posterior dislocation of the femur is a sciatic nerve palsy.

Immediately above Piriformis and more or less on a plane with it is **Gluteus Minimus.** This muscle arises from the outer surface of the ilium in the fan-shaped area enclosed between the indistinct inferior gluteal line below and the prominent anterior gluteal line above and behind. It inserts by a flat tendon into the anterior aspect of the greater trochanter of the femur. It is innervated by the *superior gluteal nerve*.

Immediately below Piriformis is the tendon of **Obturator Internus** accompanied by the **Superior** and **Inferior Gemellus.** The *nerve to Obturator Internus* arises separately from the sacral plexus (L5, S1, and S2) and passes out into the gluteal region below Piriformis (Fig. 52). It then crosses the outer aspect of the ischial spine and returns into the pelvis through the lesser sciatic foramen. It thus reaches Obturator Internus on its pelvic surface and, along the way, sends a branch to the Superior Gemellus. The Inferior Gemellus is usually supplied by the *nerve to Quadratus Femoris*.

Immediately below the tendon of Obturator Internus with its associated Gemelli is the flat but thick fleshy sheet of fibers belonging to **Quadratus Femoris.** This muscle arises from the outer aspect of the ischium just in front of the ischial tuberosity and runs laterally to the femur, where it inserts into the middle portion of the intertrochanteric crest of the femur. The *nerve to Quadratus Femoris* arises from the sacral plexus (L4, L5, and S1) and enters the gluteal region below Piriformis deep to the sciatic nerve and Obturator Internus to supply Quadratus Femoris and, on its way, the Inferior Gemellus as well.

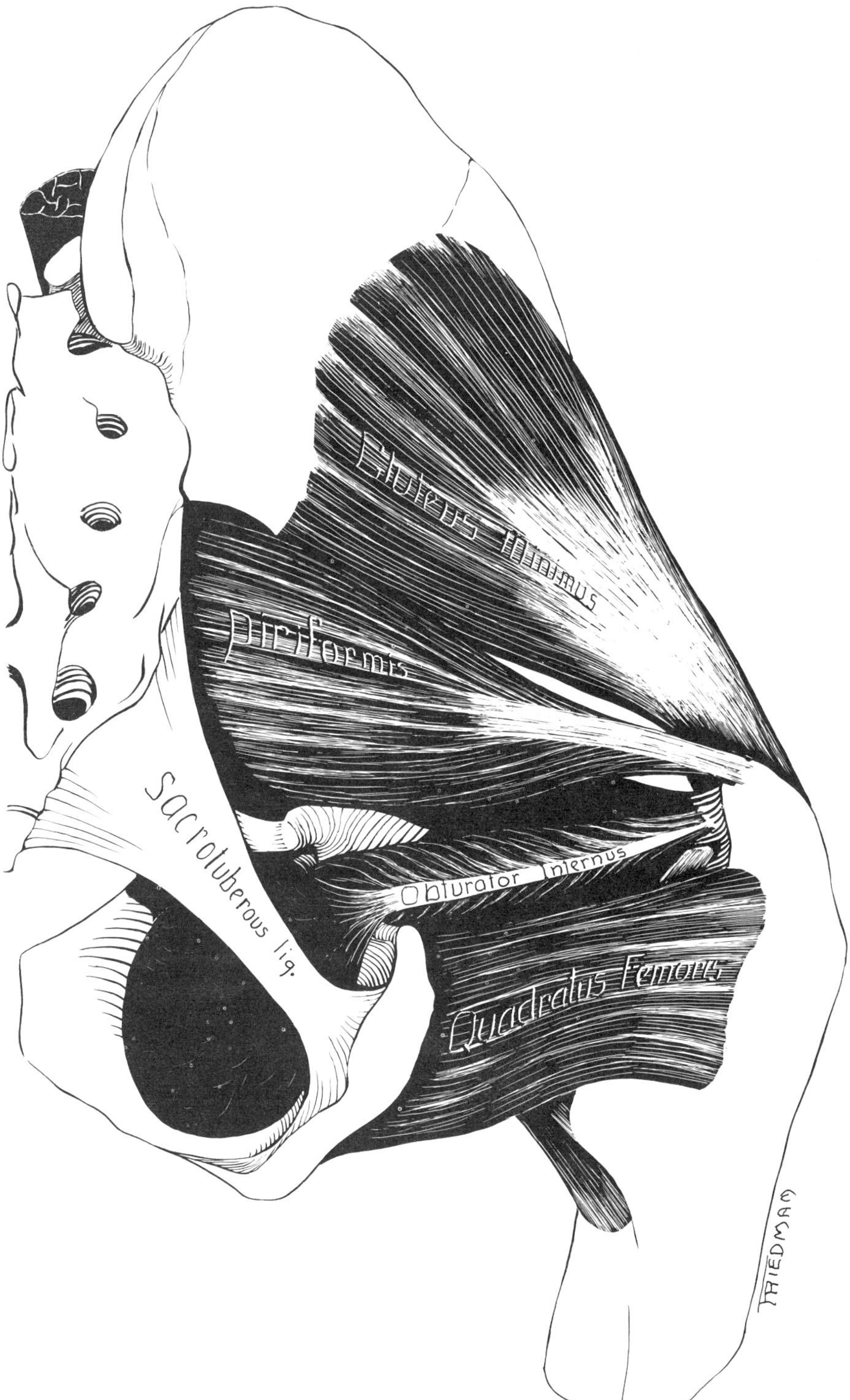
Gluteus Minimus
piriformis
Sacrotuberous lig.
Obturator Internus
Quadratus Femoris
FRIEDMAN

The Gluteal Region:
II. The Nerves and Blood Vessels

Gluteus Medius arises from the outer aspect of the iliac bone between the anterior and posterior gluteal lines. It completely covers Gluteus Minimus and, lying as it does above Piriformis, is the highest muscle seen upon reflecting Gluteus Maximus. It passes downward and laterally to insert into the flat lateral aspect of the greater trochanter. Like Gluteus Minimus, Medius is innervated by the superior gluteal nerve.

The **superior gluteal artery** and **nerve** enter the region through the greater sciatic foramen above Piriformis. They then pass laterally in the fascial interval between Gluteus Minimus and Medius and supply both muscles.

Three structures, the **pudendal nerve,** the **internal pudendal artery,** and the **nerve to Obturator Internus,** in that order from medial to lateral, enter the gluteal region below Piriformis, cross the spine of the ischium, and at once leave the gluteal region by passing forward through the lesser sciatic foramen into the perineum. The pudendal nerve and internal pudendal artery pass downward along the inferior ramus of the ischium in a fascial compartment, the *pudendal (Alcock's) canal,* which forms part of the lateral wall of the ischiorectal fossa.

The **inferior gluteal artery** and its associated **vein** also enter the gluteal region below Piriformis but divide almost at once. Branches from these vessels run to supply Piriformis, the Gemelli, and Quadratus Femoris, but by far their most important task is to endow the huge Gluteus Maximus with the requisite blood supply. The inferior gluteal artery also constantly sends descending branches which participate in the anastomosis around the back of the greater trochanter, the *cruciate anastomosis*.

The **inferior gluteal nerve** enters the region alongside the inferior gluteal vessels below Piriformis. It breaks up at once into multiple branches which penetrate the substance of the overlying Gluteus Maximus.

The **sciatic nerve,** the largest single nerve in the body, is the most prominent structure entering the gluteal region below Piriformis. It runs downward, crossing Obturator Internus and the Gemelli and then Quadratus Femoris to reach the posterior aspect of Adductor Magnus, on which it is conducted down the back of the thigh. Closely related to the sciatic nerve is the *posterior cutaneous nerve of the thigh,* which at first lies medial to it but then gradually comes to lie on its posterior aspect.

Nerves to Quadratus Femoris and **to the Gemelli** complete the roster of structures entering the gluteal region below Piriformis (Fig. 51).

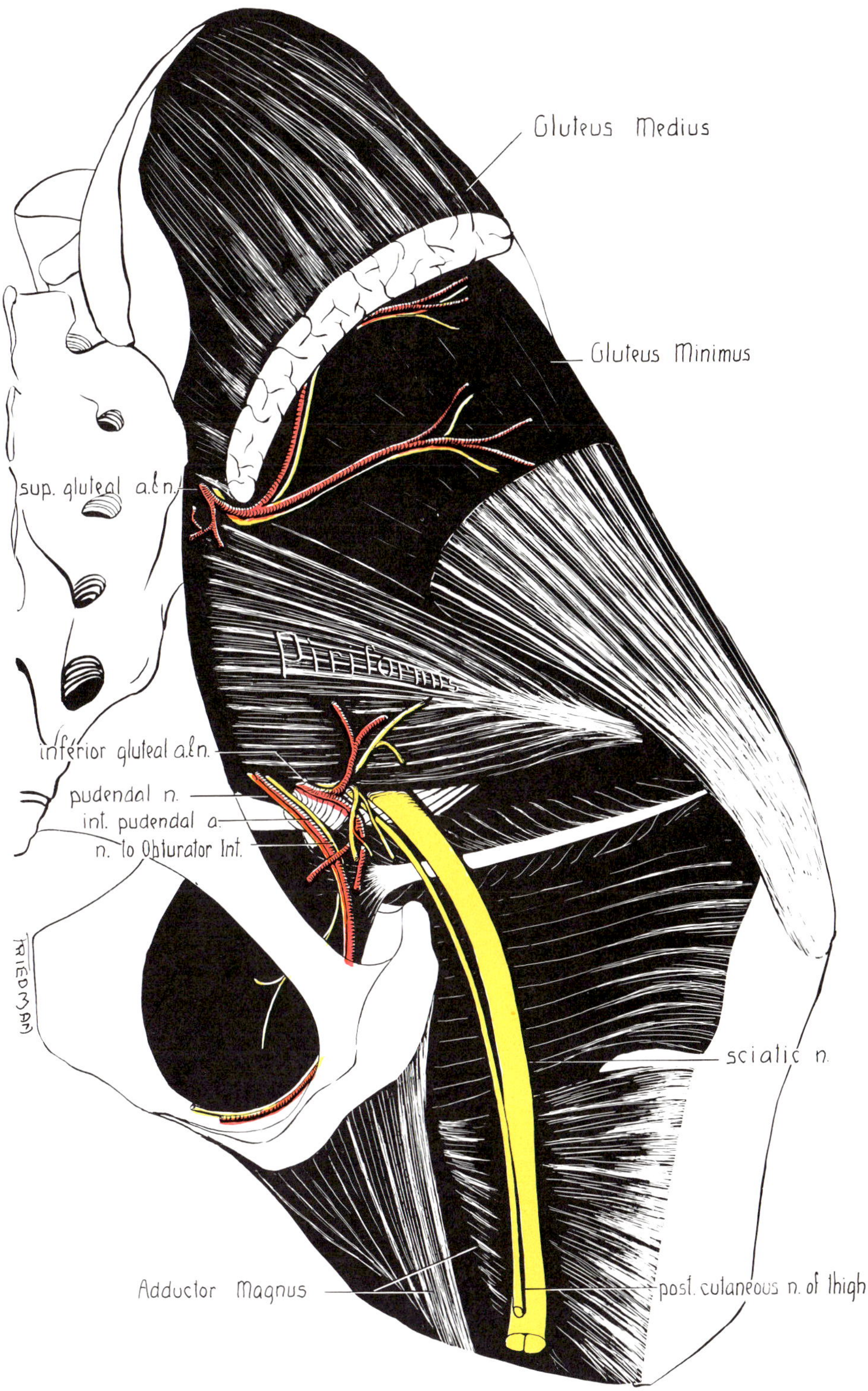
Gluteus Medius
Gluteus Minimus
sup. gluteal a.&n.
Piriformis
inferior gluteal a.&n.
pudendal n.
int. pudendal a.
n. to Obturator Int.
FRIEDMAN
sciatic n.
Adductor Magnus
post. cutaneous n. of thigh

The Gluteal Region:
III. Gluteus Maximus

Completely covering all of the structures of the gluteal region is the massive **Gluteus Maximus.** It arises from a small area of the iliac bone medial to the posterior gluteal line and from the adjacent margin of the sacrum. In addition to these bony points of anchor, the muscle also arises extensively from dense fascia overlying the back of the sacrum and extending upward to anchor to the outer margin of the iliac crest. Laterally, the upper two thirds of the muscle inserts into the dense fascia of the iliotibial tract which overlies the greater trochanter. The lower third inserts into a rough gluteal impression on the back of the femur. The muscle presents a free lower border which sometimes can be palpated. The lower fibers overlie the ischial tuberosity as well as the hamstring muscles which originate from the tuberosity. The inferior border of Gluteus Maximus is so placed that the ischial tuberosity and its covering bursa form the surface for sitting, leaving the lowest muscular fibers to ride free.

The great muscle masses of the gluteal region are commonly selected as depot for the intramuscular injection of therapeutic agents. The upper lateral quadrant of the region provides the least opportunity for misadventure with the large nerves, arteries, and veins of the region. By the same token, the lower medial quadrant is the most hazardous and is never used.

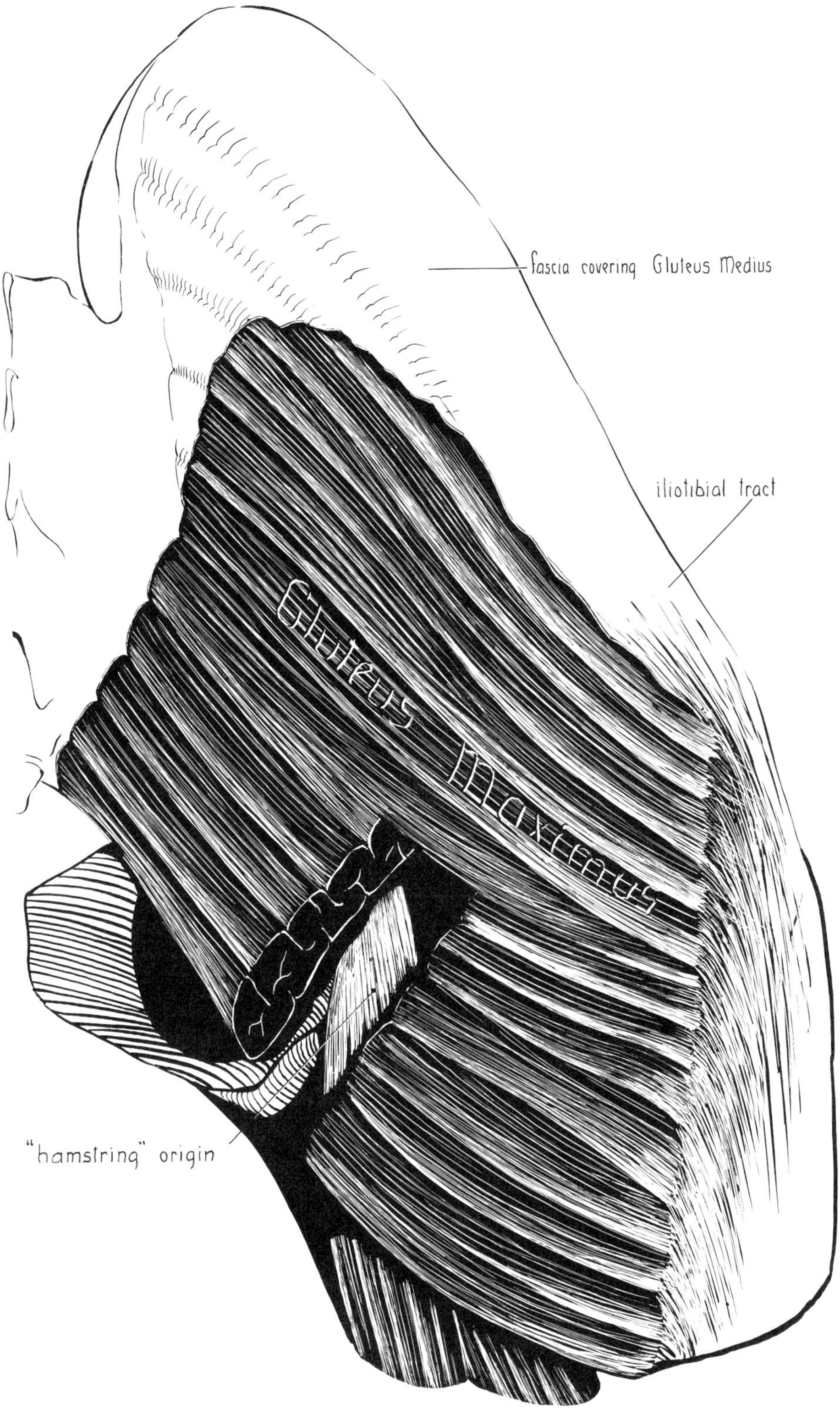
fascia covering Gluteus Medius
iliotibial tract
Gluteus Maximus
"hamstring" origin

The Front of the Thigh

The Deep Adductor Plane: Adductor Magnus and Gracilis

Adductor Magnus is a great, fan-shaped muscle which extends laterally and downward from the inferior ramus of the pubis and ischium to insert over almost the full length of the femur. It forms a natural dividing plane between the front and back of the thigh. Quadratus Femoris lies above it on the same plane, while its medial border is continuous with Gracilis.

It should be appreciated that the origin of Adductor Magnus lies roughly in the horizontal plane, while its insertion is essentially vertical. This means that the muscle must undergo almost 90 degrees of rotation. This it does close to its origin, and as this illustration attempts to show, those fibers that arise most anteriorly run almost horizontally while those that arise most posteriorly (close to the ischial tuberosity) run almost vertically downward. In fact, these vertical fibers which arise close to the ischial tuberosity belong to the hamstring group and are so innervated, as will be described later. (This twisting of the muscle fibers is seen from behind in Figure 65.) Most of Adductor Magnus inserts linearly beside the lateral lip of the linea aspera along the full length of the back of the femur. The most medial vertical portion of the muscle detaches to form a separate tendon, which inserts into the adductor tubercle just above the medial epicondyle of the femur.

The adductor tubercle and the tendon of insertion of Adductor Magnus are easily palpated and are useful as surface references. The separation of the two components of Adductor Magnus leaves a triangular interval between their inserting fibers which allows the femoral vessels to pass from the front to the back of the thigh.

The intermuscular interval between Adductor Magnus and Quadratus Femoris provides a convenient passageway for the *medial circumflex femoral artery,* which runs from the femoral artery to join the inferior gluteal artery and its branches in the *cruciate anastomosis* around the back of the greater trochanter.

Gracilis takes origin from a vertical line alongside the inferior ramus of the pubis just medial to Adductor Magnus, and runs down as a thin, straplike muscle to form a tendon. This inserts, as will be seen in Figure 63, in the upper portion of the medial aspect of the tibia. Gracilis is easily palpated as the most medial muscle of the upper portion of the thigh.

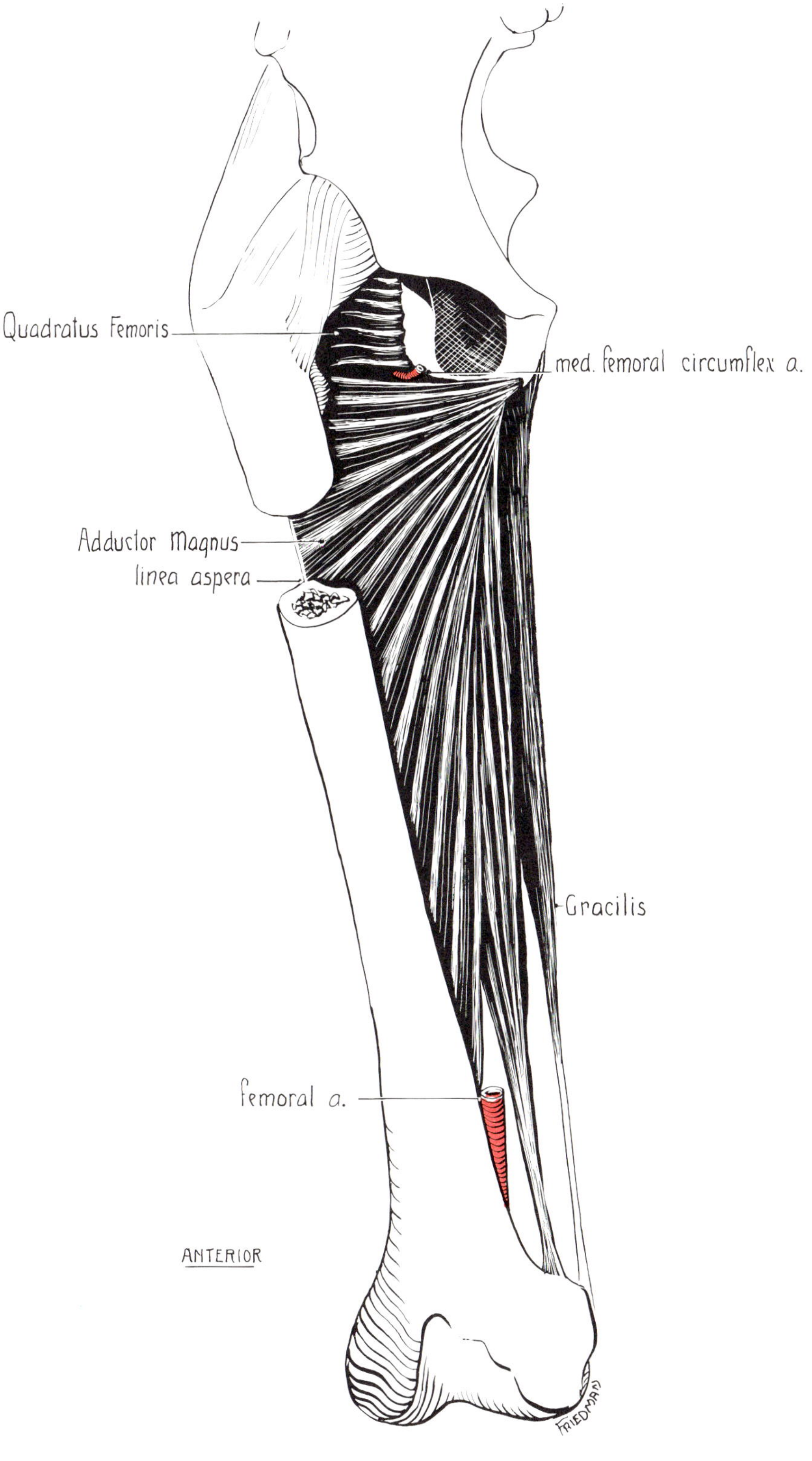
Quadratus Femoris
med. femoral circumflex a.
Adductor Magnus
linea aspera
Gracilis
femoral a.
ANTERIOR
FRIEDMAN

The Middle Adductor Plane: Obturator Externus and Adductor Brevis

Two muscles partially occupy the plane immediately in front of Adductor Magnus. The higher of these, **Obturator Externus,** has already been briefly described in relation to the capsule of the hip joint. It arises from the obturator membrane and from its immediately adjacent bony margin. The fleshy fibers of origin taper rapidly to form a tendon which passes first below and then behind the capsule of the joint to insert into a pit located in the shadow of the intertrochanteric crest.

Adductor Brevis arises from the inferior ramus of the pubis below the pubic tubercle and rather parallel to the symphysis pubis. It fans out as a flat but thick and fleshy muscle to insert into the middle of the linea aspera posteriorly. It covers approximately half of Adductor Magnus.

The **obturator nerve** leaves the pelvis through the obturator canal and enters the thigh on, often partly through, Obturator Externus. In the thigh it divides into two major components, an anterior (superficial) and a posterior (deep) branch. The *anterior branch* continues down on Adductor Brevis while the *posterior branch* passes between Adductor Brevis and Adductor Magnus and continues below Adductor Brevis to run with the femoral vessels to the back of the knee. The obturator nerve innervates all of the adductor muscles except for those fibers of Adductor Magnus of ischial origin that are, like the hamstrings, supplied by the sciatic nerve. For the most part, the anterior branch of the obturator nerve innervates the more superficial planes of muscle, the posterior branch the deepest plane.

In the preceding illustration, the *medial circumflex femoral artery* was seen to pass to the back of the thigh between Quadratus Femoris and Adductor Magnus. Here, in the same relative position, it is seen passing below Obturator Externus.

The **profunda femoris artery** spends most of its course on Adductor Brevis and terminates by continuing for a short distance below this on Adductor Magnus. It gives off at least four major *perforating branches* which pass to the back of the thigh. The highest two of these of necessity pierce both Adductor Brevis and Adductor Magnus to reach their final distribution; the third and the fourth, which is the termination of the profunda femoris, need only pierce Adductor Magnus to accomplish the same end.

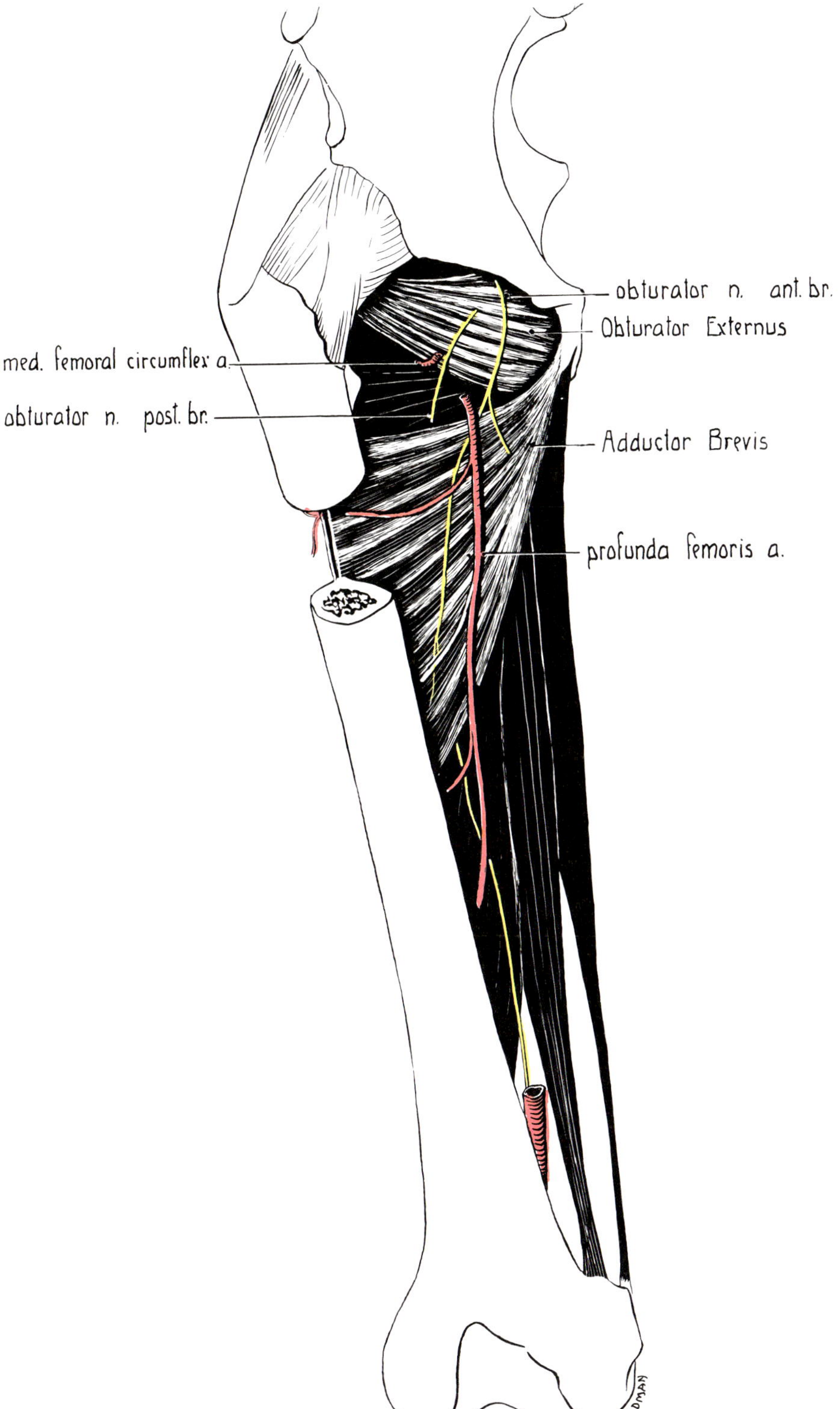
obturator n. ant. br.
Obturator Externus
med. femoral circumflex a.
obturator n. post. br.
Adductor Brevis
profunda femoris a.
RIEDMAN

The Superficial Adductor Plane: Iliopsoas, Pectineus, and Adductor Longus

Both Psoas and Iliacus arise as abdominal muscles. **Psoas** originates from alongside the lumbar vertebrae and forms a well-rounded muscle mass which passes along the lateral margin of the pelvic brim. It is joined on its lateral side by **Iliacus,** which arises as a flat muscle from the concavity of the iliac fossa. As both muscles, now blended, pass behind the inguinal ligament, they become flat and tendinous, angled one to another so as to form a rather shallow V-shaped trough. This conjoined tendon of **Iliopsoas** passes in front of and then below the capsule of the joint to insert into the lesser trochanter of the femur and into the bone below this for a short distance (see also Figure 50).

Pectineus arises from the upper surface of the pubis, actually from its pectineal surface in front of the *pectineal line.* From here it runs down and laterally as a flat sheet and inserts into the medial lip of the linea aspera below the Iliopsoas tendon.

Adductor Longus arises adjacent to Pectineus, actually from the body of the pubis just lateral to the symphysis pubis. It passes down as a flat sheet of muscle on the same plane with Pectineus to insert into the medial lip of the linea aspera immediately below it.

It is clear then that Iliopsoas, Pectineus, and Adductor Longus form a continuous sheet within the thigh. Iliopsoas, however, is not truly an adductor, but is rather a flexor of the femur. Its innervation (L2, 3, 4) reaches it in part directly from the lumbar plexus and in part from the femoral nerve. Pectineus receives its nerve supply like an adductor muscle from the obturator nerve through its anterior branch and frequently receives additional innervation through the femoral nerve. Adductor Longus is wholly innervated as an adductor, that is, from the anterior branch of the obturator nerve.

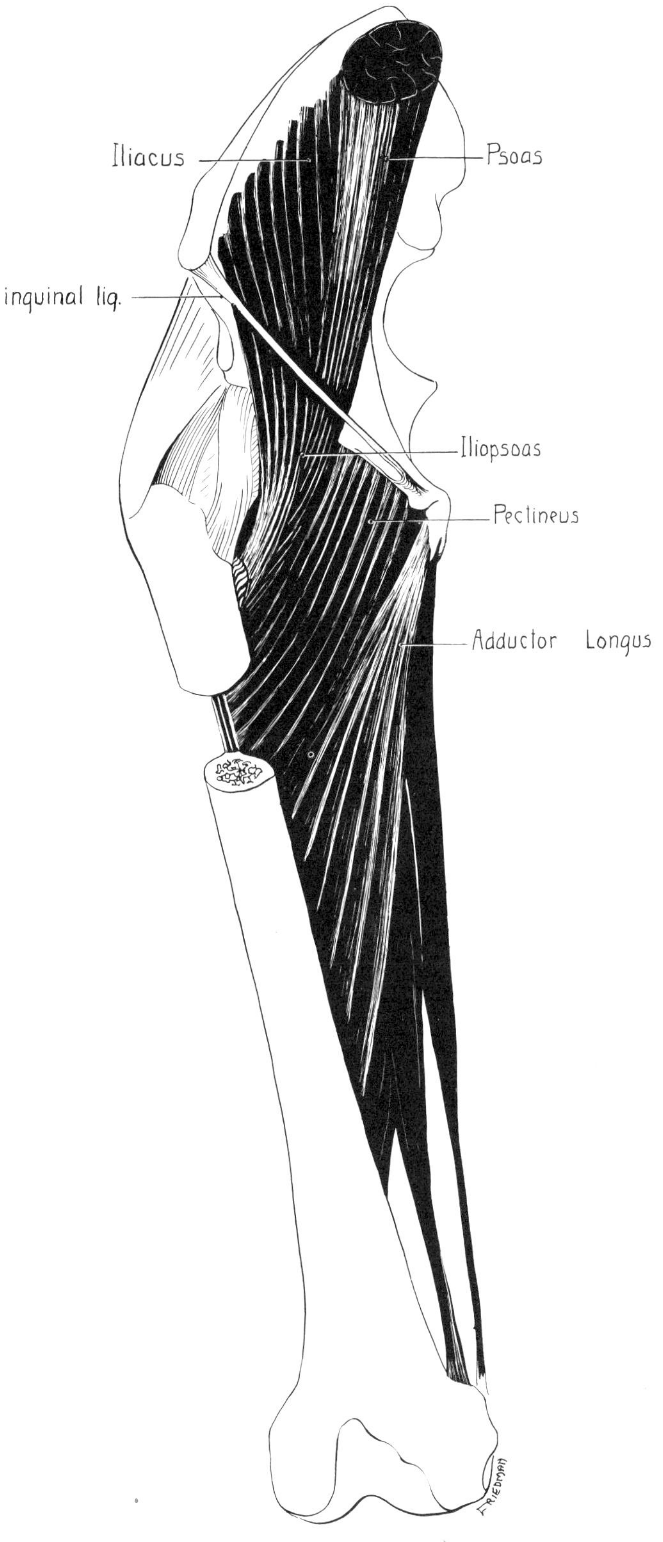

Iliacus
Psoas
inguinal lig.
Iliopsoas
Pectineus
Adductor Longus
FRIEDMAN

The Femoral Nerve, Artery, and Vein

The **femoral artery** and **vein** enter the thigh as the continuations of the external iliac vessels enclosed in a fascial sleeve derived from the endoabdominal fascia, the *femoral sheath.* This sheath extends below the inguinal ligament for 2 or 3 cm and houses the femoral artery most laterally and the femoral vein in its middle portion, while its most medial portion, the *femoral canal,* remains essentially empty, filled only with a lymph node or two and some fat. The femoral canal is the natural pathway for femoral hernias.

The femoral artery and vein run almost vertically down through the thigh. Because of the obliquity of the femur, they are at first placed at some remove from the bone, but gradually approach it inferiorly as they pass through the hiatus in Adductor Magnus to reach the back of the femur in the popliteal fossa. The vein, at first medial to the artery, becomes progressively more posterior. It is thus the more superficial structure in the popliteal fossa when this is approached surgically from behind.

The **femoral nerve** is formed by large branches of L2, 3, and 4 in the lower abdomen, in the interspace between Psoas and Quadratus Lumborum. It continues obliquely downward in this plane and crosses the iliac fossa lying on Iliacus and overlapped by Psoas. As these muscles become tendinous and hence narrower, the femoral nerve comes fully into view and, as it passes into the thigh behind the inguinal ligament, lies in the trough between Iliopsoas and Pectineus. Now flat and broad, it breaks up almost at once into its many branches. Its *muscular branches* all run to the components of the Quadriceps muscle, to Sartorius and on occasion, to Pectineus as well. In addition, it gives rise to numerous cutaneous branches described as *lateral, intermediate,* and *medial.* One of the many branches of the femoral, the *saphenous nerve,* is usually considered its termination. This nerve descends through the thigh along with the femoral vessels. It lies at first lateral to the artery, but then crosses in front of it to the medial side at the lower third of the thigh. From here it runs directly to the medial side of the knee and ends in the skin of the medial side of the foot.

The **profunda femoris artery** arises from the femoral before it leaves Pectineus. It separates from the parent vessel at the lower border of the Pectineus muscle to descend between Adductor Longus and Adductor Brevis (Fig. 55). Continuing below Adductor Brevis on Adductor Magnus it ends as the *fourth perforating artery.*

The *medial circumflex femoral artery* arises from the profunda femoris or often directly from the femoral artery itself. It then dips backward to run first between Iliopsoas and Pectineus, then between Obturator Externus above and Adductor Brevis below, and finally enters the region of the cruciate anastomosis on the back of the femur by passing between Quadratus Femoris and Adductor Magnus. The *lateral circumflex femoral artery* also arises from the profunda or from the femoral and runs obliquely down and laterally to the Quadriceps. In addition it sends transverse branches around the femur just below the greater trochanter; these also participate in the cruciate anastomosis.

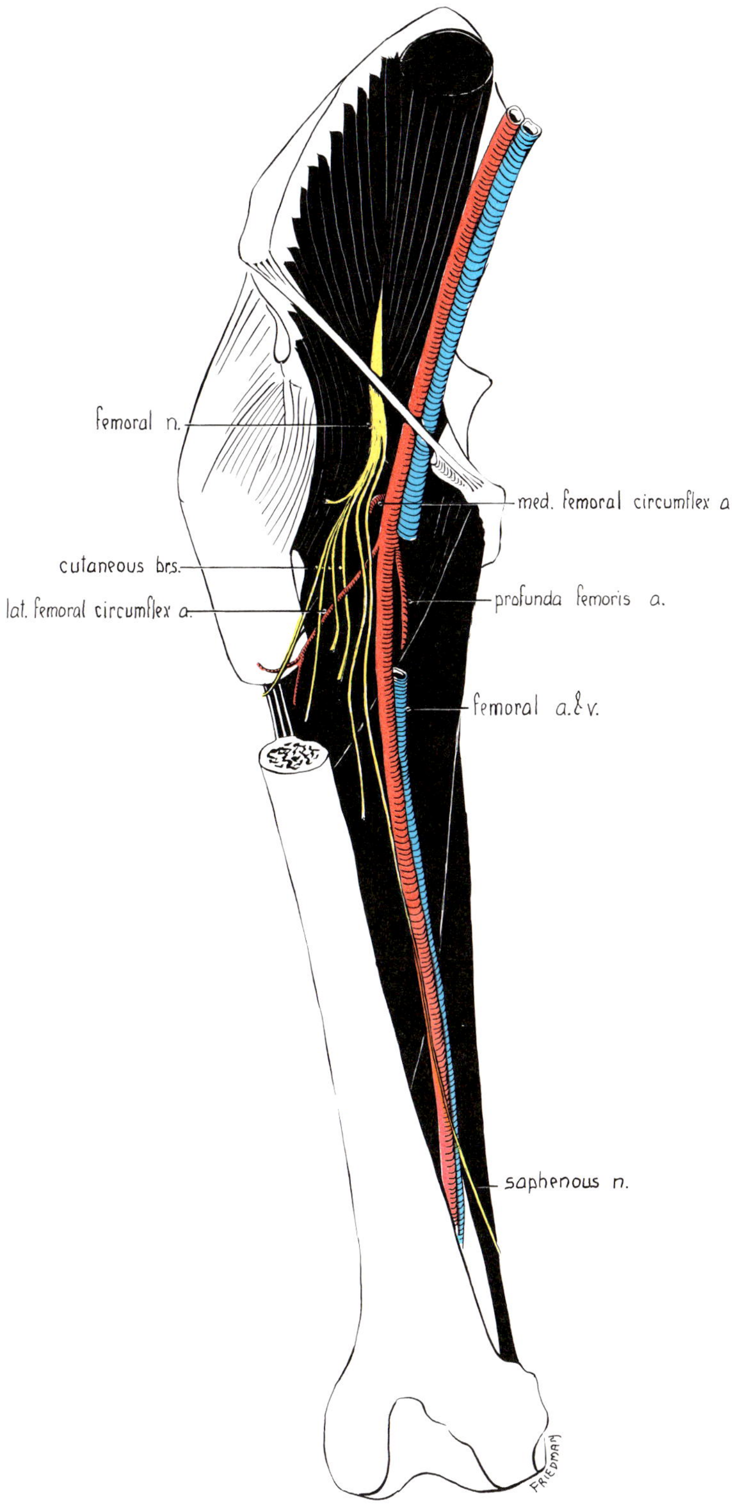
femoral n.
med. femoral circumflex a.
cutaneous brs.
profunda femoris a.
lat. femoral circumflex a.
femoral a.&v.
saphenous n.
FRIEDMAN

Quadriceps Femoris and Sartorius

The adductor muscles occupy the interval between the lower half of the hip bone and the femur. Since the femur is set obliquely, this interval is triangular—wide above, and tapering to its apex below. The Quadriceps group of muscles, on the other hand, follows precisely the line of the femur and forms a cylinder of muscle surrounding the bone. All of the shaft of the femur is preempted by the components of this muscle group, only the linea aspera posteriorly remaining free for the attachment of the adductors. The fascial enclosure occupied by the Quadriceps is fundamentally separate from that of the adductors, although where the former directly overlies the latter, these fascias blend.

Sartorius is a straplike muscle which passes diagonally down the front of the thigh marking out the trough between the adductor group and the medial border of the Quadriceps. The femoral vessels are overlapped from their lateral side by the Quadriceps and covered anteromedially by Sartorius as they lie in front of Adductor Magnus in the lower part of their course. The rather triangular enclosure so formed is termed the *adductor (subsartorial) canal.*

The dense fascia (fascia lata) enclosing the muscles of the thigh is reinforced by a dense ligamentous band which passes down as a ribbon 5 cm or more wide along the lateral side. The upper portion of this *iliotibial tract* is supported by **Tensor Fasciae Latae,** a short but stout muscle which hooks it to the anterior quarter of the lateral lip of the iliac crest. The lower end of the tract passes into the lateral condyle of the tibia. It will be recalled that the posterior portion of the iliotibial tract also receives fibers of insertion from Gluteus Maximus (Fig. 53). Tensor Fasciae Latae, being fundamentally a portion of the gluteal musculature, is innervated by the *superior gluteal nerve.*

Vastus Intermedius is the most centrally placed part of the Quadriceps group and closely embraces the lateral and anterior surfaces of the femur. It is completely covered by the other members of the group. **Vastus Lateralis** has a linear origin which extends from the intertrochanteric line around the lateral side of the femur to the lateral lip of the linea aspera. It wraps itself around Vastus Intermedius but can be partially separated from it by retraction. **Vastus Medialis** originates from the medial portion of the shaft of the femur above, alongside, and below the medial lip of the linea aspera. It covers but is not attached to the medial side of the femur. These three muscles extend upward as far as the greater trochanter and the intertrochanteric line.

Rectus Femoris, the fourth muscle of the group, arises separately from the others, from the anterior inferior iliac spine and from the bone directly above the acetabulum. The flat tendon of origin passes down in front of the capsule of the hip joint before giving rise to the fleshy muscle belly which runs down the length of the femur in front of Vastus Intermedius. All four muscles become tendinous inferiorly where they insert into the superior and medial borders of the patella and continue via the *ligamentum patellae* into a large tuberosity on the anterior aspect of the tibia. All are innervated by the femoral nerve.

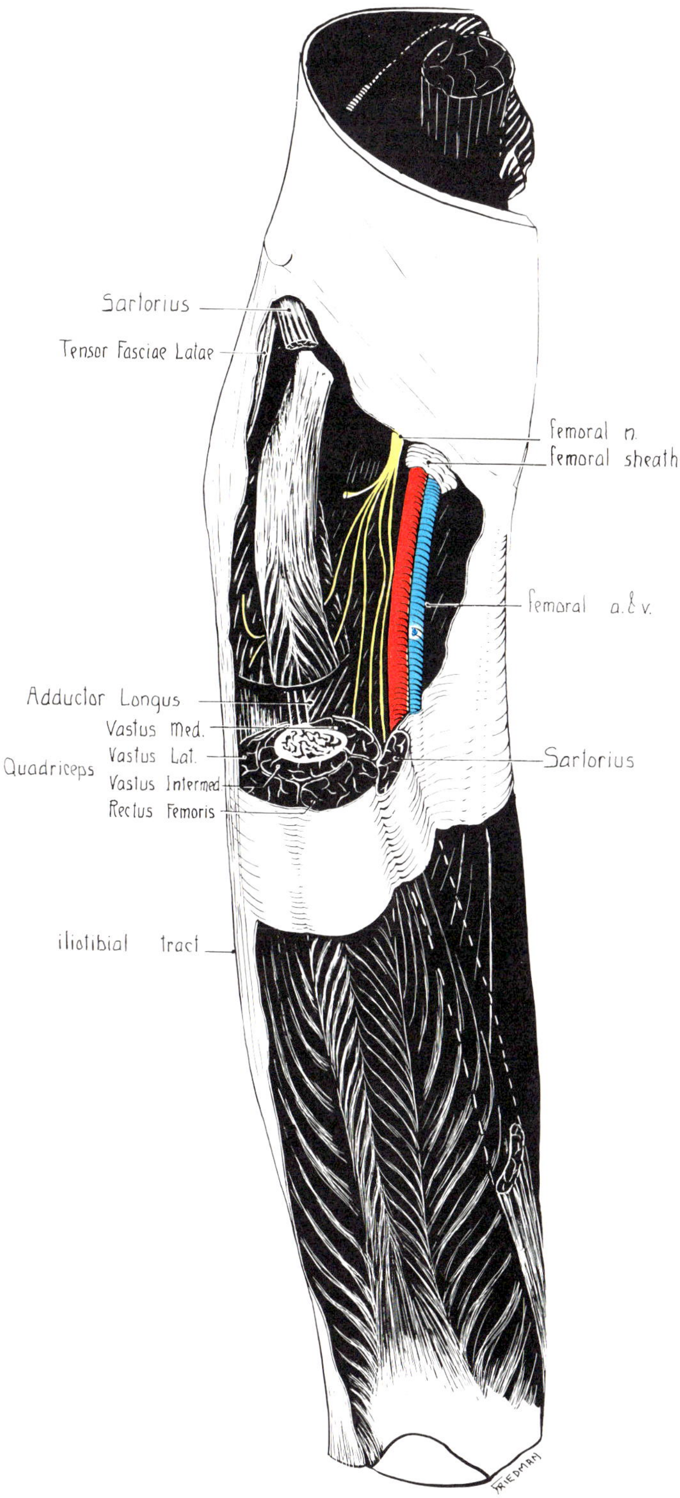
Sartorius
Tensor Fasciae Latae
femoral n.
femoral sheath
femoral a. & v.
Adductor Longus
Vastus Med.
Quadriceps
Vastus Lat.
Vastus Intermed
Rectus Femoris
Sartorius
iliotibial tract
FRIEDMAN

The Great Saphenous Vein

Some of the veins of the body have special clinical significance. The **great saphenous vein** is one of these. The venous drainage of the lower limb will be considered as a whole in Figure 83, but we should note at this point that a large proportion of the venous drainage of the foot and leg is carried by superficial veins. Part of this drainage joins deep channels at the popliteal fossa. The remainder, for the most part, continues up the thigh in the saphenous vein to unite with the deep venous system just below the inguinal ligament. There are additionally a few perforating veins in the calf and thigh which allow flow from the great saphenous vein to the deep veins.

The great saphenous vein is reminiscent of the cephalic vein in the upper limb. It arises in the dorsal venous plexus on the foot and passes up around the medial side of the leg. It then runs on the medial side of the knee superficial to the fascia covering Sartorius and moves slowly anteriorly as it runs up the front of the thigh to reach the *saphenous opening* (*fossa ovalis*), a crescentic aperture in the fascia lata 2 or 3 cm below the inguinal ligament. On its way, it lies free in the subcutaneous fat and receives numerous tributary veins not only from the thigh but from the lower part of the abdominal wall as well.

The superficial veins of the leg are in large part unsupported and are prone to become tortuous, enlarged, and ultimately varicose. The saphenous vein is a conventional point of surgical attack for procedures designed to deal with these problems.

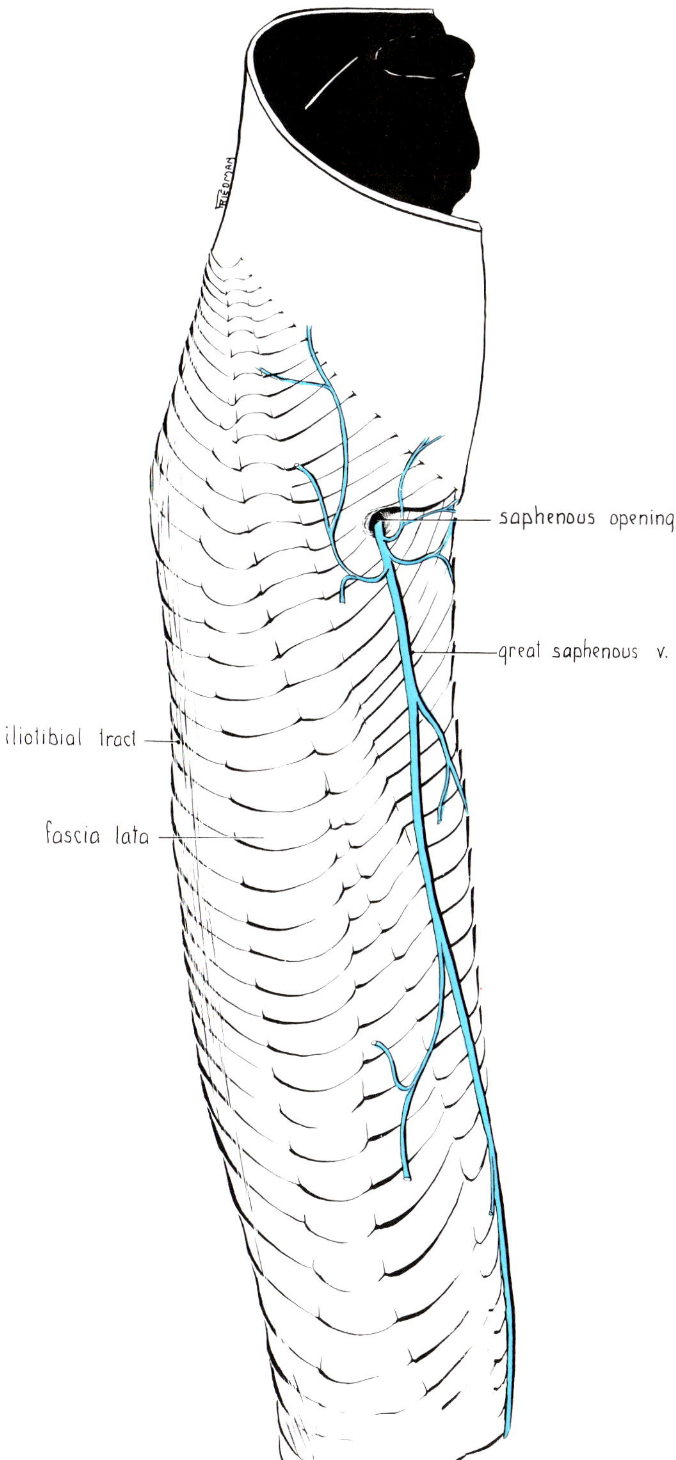
saphenous opening
great saphenous v.
iliotibial tract
fascia lata

The Knee Joint

THE KNEE JOINT

The Articulatory Surfaces

It is worth emphasizing that the front of the knee is homologous with the back of the elbow, just as the cubital fossa at the front of the elbow recalls the popliteal fossa at the back of the knee. It is the rotation of the lower limb as it descends during development in preparation for the upright posture that is responsible for the reversal in relative position. Anatomical terminology, however, respects the basic homology of the upper and lower limb and so considers extension of the knee a forward movement of the leg, and flexion a backward movement. Similarly, the anterior aspect of the leg and its continuation, the dorsum of the foot, are extensor in function parallel with the back of the forearm and the back of the hand.

The knee joint is a simple hinge which permits mainly flexion and extension but possesses, in addition, a small rotatory component of some importance. Despite the very great loads it must support (e.g., running, jumping), the joint is not stabilized by bony buttresses, as is the homologous elbow joint, but depends instead upon its ligamentous arrangements.

The weight-bearing components of the knee joint ride on the plateau of the tibia. This consists of two massive *condyles* which support shallow-dished lateral and medial articular surfaces. These are separated by a contorted *intercondylar eminence,* which obviously provides anchor points for knee joint ligaments. The articular surfaces are moderately deepened by fibrocartilaginous *menisci,* each triangular in cross section and half moon in outline. The free ends, or horns, of each meniscus are firmly anchored to bone, leaving the rest of the meniscus to adapt to movement. The lateral meniscus is considerably smaller than the medial, as is its supporting condyle, and its horns attach within the horns of the medial meniscus. A ligamentous band runs transversely between the two menisci anteriorly.

The larger size of the medial bearing surface relative to the lateral corresponds to the relative sizes of the femoral condyles and, in addition, permits the medial condyle of the femur to slide slightly during extension. Thus, in extension, the medial condyle of the femur slides backward, pivoting around the lateral condyle. The sliding action gradually brings the flat anterior portion of the medial condyle of the femur into direct apposition with the tibia, so that when the knee is fully extended the weight of the body is supported on a rather flat table. This minimizes the amount of energy required to maintain the fully upright posture and is the knee locking mechanism. Flexion reverses this and begins with a forward sliding of the medial condyle of the femur, which brings the rounded portion of the condyle into position before the hinge motion is actually initiated.

The apposition of the femur and the tibia depends upon the cruciate ligaments. The **anterior cruciate ligament** arises from the intercondylar area between the horns of the menisci. It passes upward and laterally to insert into the medial aspect of the lateral condyle of the femur. The **posterior cruciate ligament** runs from the fossa immediately behind the intercondylar eminence forward and upward to the lateral aspect of the medial condyle of the femur.

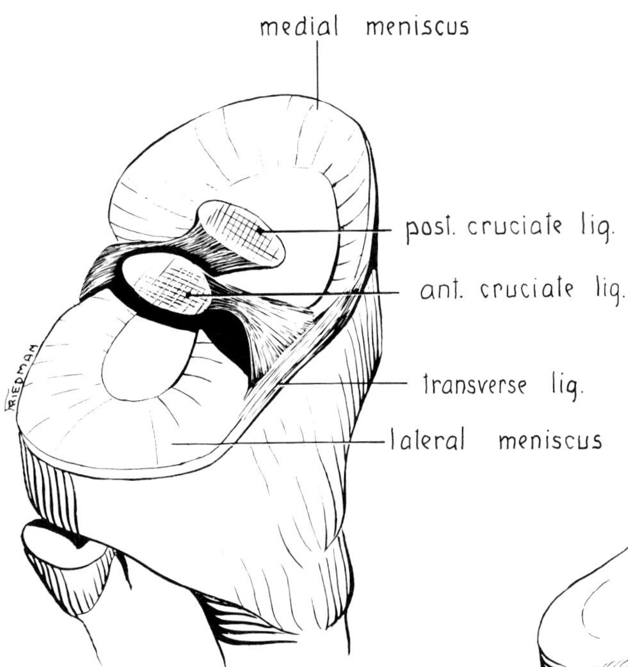
medial meniscus
post. cruciate lig.
ant. cruciate lig.
transverse lig.
lateral meniscus
FRIEDMAN

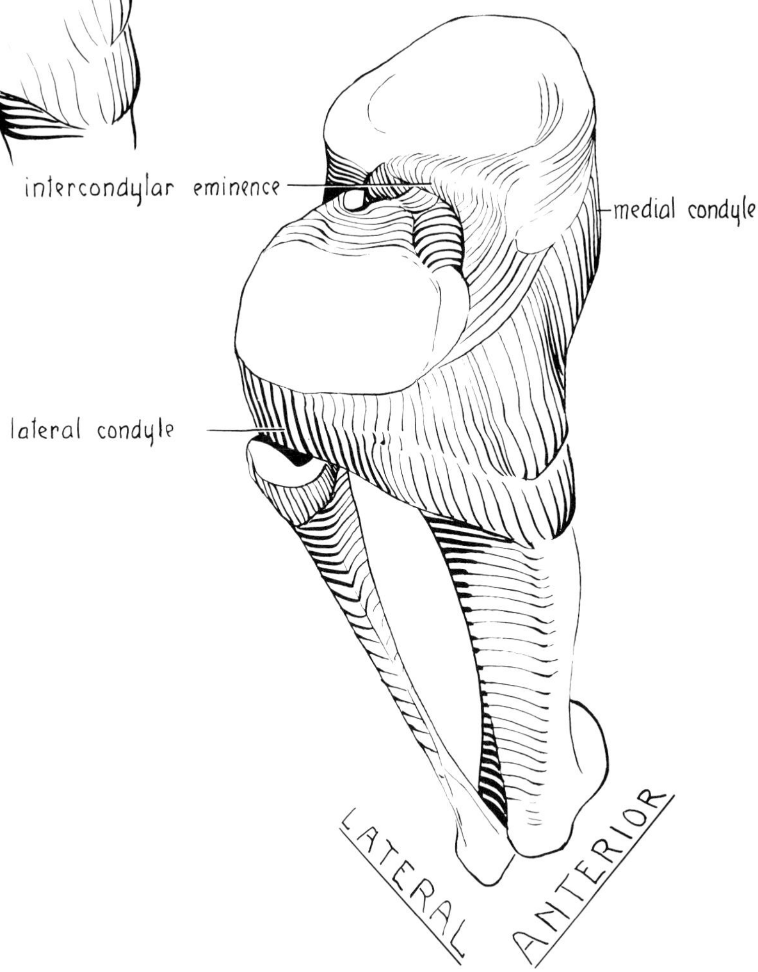
intercondylar eminence
medial condyle
lateral condyle
LATERAL
ANTERIOR

The Synovial Membrane

The **synovial membrane** attaches inferiorly to the upper margin of each meniscus, leaving the cartilage surface bare. It is reflected around the sides and anterior aspect of the cruciate ligaments so that these are effectively excluded from the joint cavity. This portion of the membrane passes upward to the articular margins of each femoral condyle. Right and left sides of the joint space communicate anteriorly above and below a ridge of synovial membrane. This ridge extends as a single conical *infrapatellar fold* in front of the cruciate ligaments and spreads laterally to form two fringed *alar folds* as it passes forward to the lower margin of the patella.

The anterior wall of the synovial capsule extends high up on the femur for about a hand's breadth above the condyles, then, recurving, lines the face of the bone down to the upper margin of its articular surfaces. The patella is, as it were, embedded in this part of the synovial membrane.

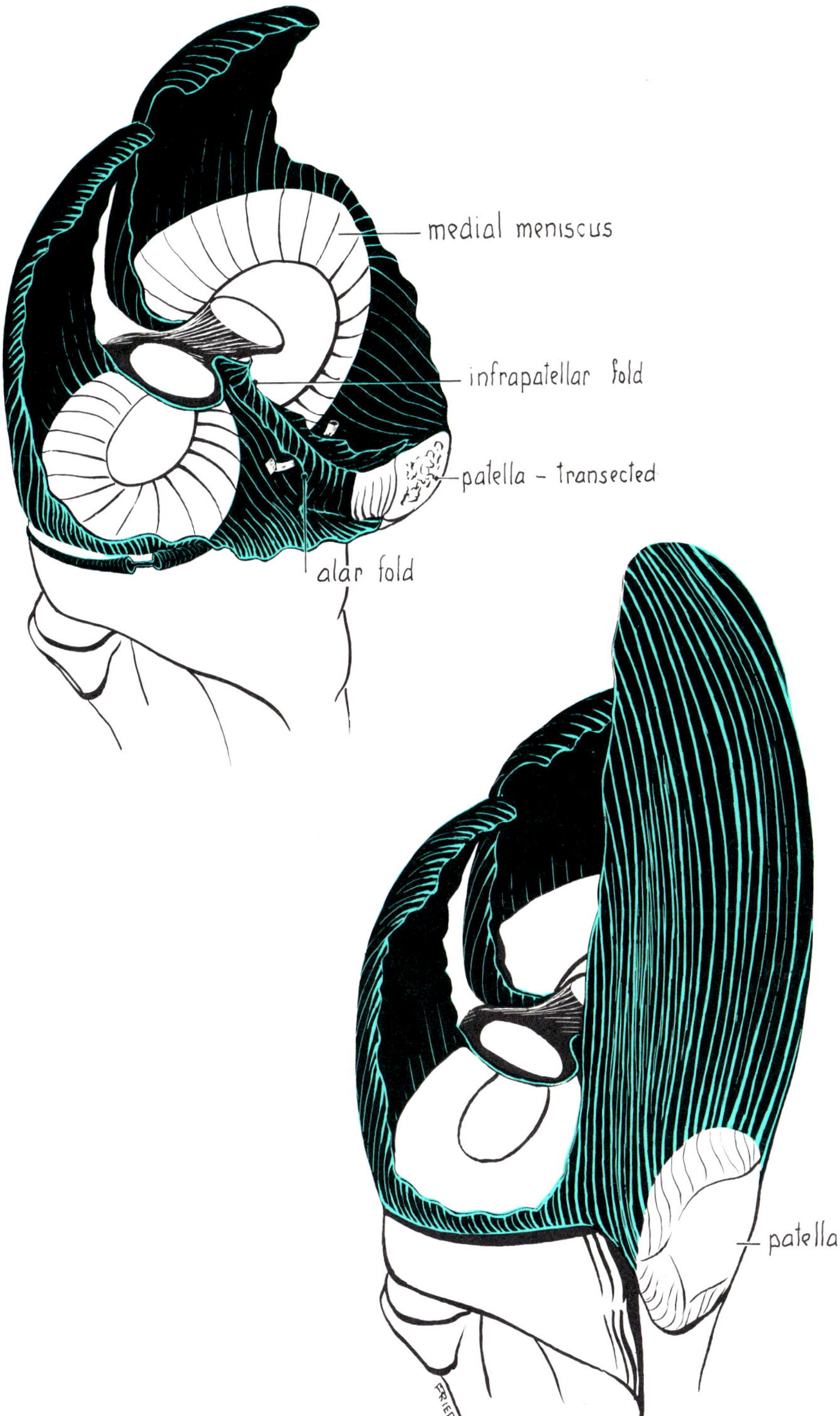
medial meniscus
infrapatellar fold
patella - transected
alar fold
patella
FRIEDMAN

The Fibrous Capsule

In this illustration the knee joint has been incised longitudinally in the midline and the two halves retracted to expose the interior of the joint and, in particular, the anterior extent of the synovial membrane. Inferiorly, the infrapatellar fold has been incised so that the inferior portion of the synovial membrane between meniscus and patella is displayed in section. Above the patella the synovial membrane forms a sac, which is reflected down on the anterior and lateral aspects of the femur as far as the articular margin. This upper part of the joint cavity is, in fact, a bursa between the quadriceps tendon and the bone. This *suprapatellar* (*Quadriceps*) *bursa* is large and not sharply demarcated from the joint cavity with which it communicates.

The anterior portion of the **fibrous capsule** of the knee joint is the tendinous component of the insertion of Quadriceps. Superiorly, it is formed for the most part by Vastus Intermedius and, overlying this, the tendon of insertion of Rectus Femoris, with which it fuses before reaching the patella. Laterally, both above and at the sides of the patella, the capsule is formed by Vastus Lateralis; medially by Vastus Medialis. All of these fibers are consolidated below the patella as the *ligamentum patellae,* which inserts into the *tibial tuberosity.*

In addition to the suprapatellar bursa, two other bursae at the front of the knee are of clinical importance. One of these is the *deep infrapatellar bursa,* which lies between the tibia and the ligamentum patellae. The other, the *prepatellar bursa,* lies between the skin and the lower part of the patella and adjacent ligamentum patellae; it commonly becomes inflamed (housemaid's knee).

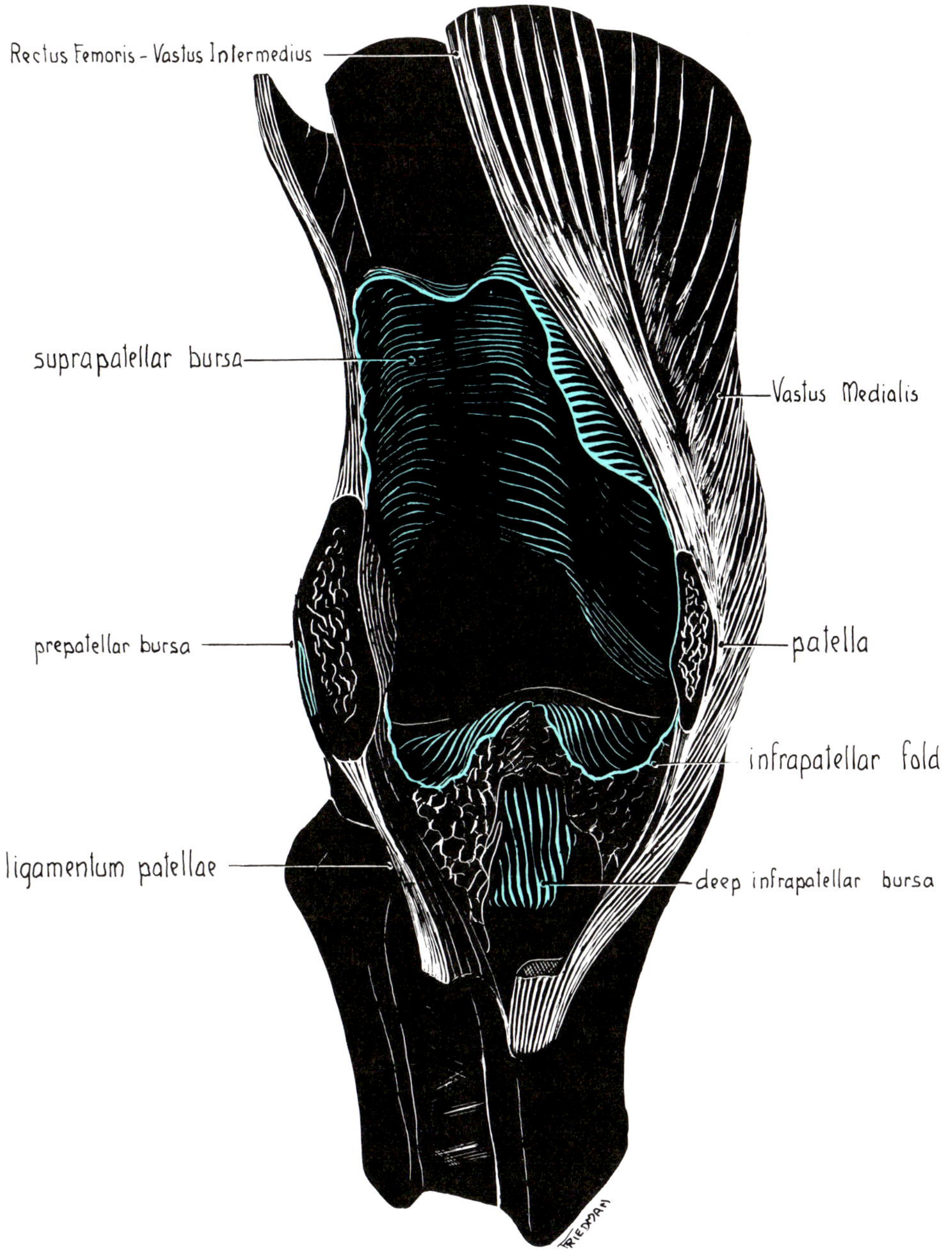
Rectus Femoris - Vastus Intermedius
suprapatellar bursa
Vastus Medialis
prepatellar bursa
patella
infrapatellar fold
ligamentum patellae
deep infrapatellar bursa
FRIEDMAN

The Medial and Lateral Stabilizing Ligaments

It is apparent that a simple hinge joint without bony flanges must be strongly buttressed on its sides by collateral ligaments. Since the femur angles outward from the horizontal plateau of the tibia, a particularly heavy strain is thrown on the medial side of the knee joint. Thus the tibial (medial) collateral ligament, in the first line of defense, must be considerably more extensive than the fibular (lateral).

The **tibial collateral ligament** is represented by two stout, flat bands. These arise in common from the medial epicondyle of the femur just below the insertion of Adductor Magnus on the adductor tubercle. The deeper band passes downward and backward to insert into the side and back of the medial condyle of the tibia. The more superficial fibers pass straight downward to be anchored in the shaft of the tibia immediately below the condyle. The tibial collateral ligament is firmly joined to the outer margin of the medial meniscus and greatly limits the ability of this cartilage to move adaptively with the femoral condyle. The medial side of the knee joint capsule also receives some further support from the overlying tendons of Sartorius, Gracilis, and Semitendinosus.

The **fibular collateral ligament** is a dense round band that extends from the middle of the lateral epicondyle of the femur straight down to the top of the head of the fibula. It is separated from the underlying lateral meniscus by the tendon of insertion of the Popliteus muscle, which passes into the femur and to a lesser degree into the meniscus as well. The lateral side of the knee joint capsule receives some further support from the iliotibial tract, which passes down to the lateral condyle of the tibia and fuses with the capsule.

The probable reason for the infrequency of a torn lateral meniscus is that Popliteus pulls it out of the way. The medial meniscus has no similar protection, of course.

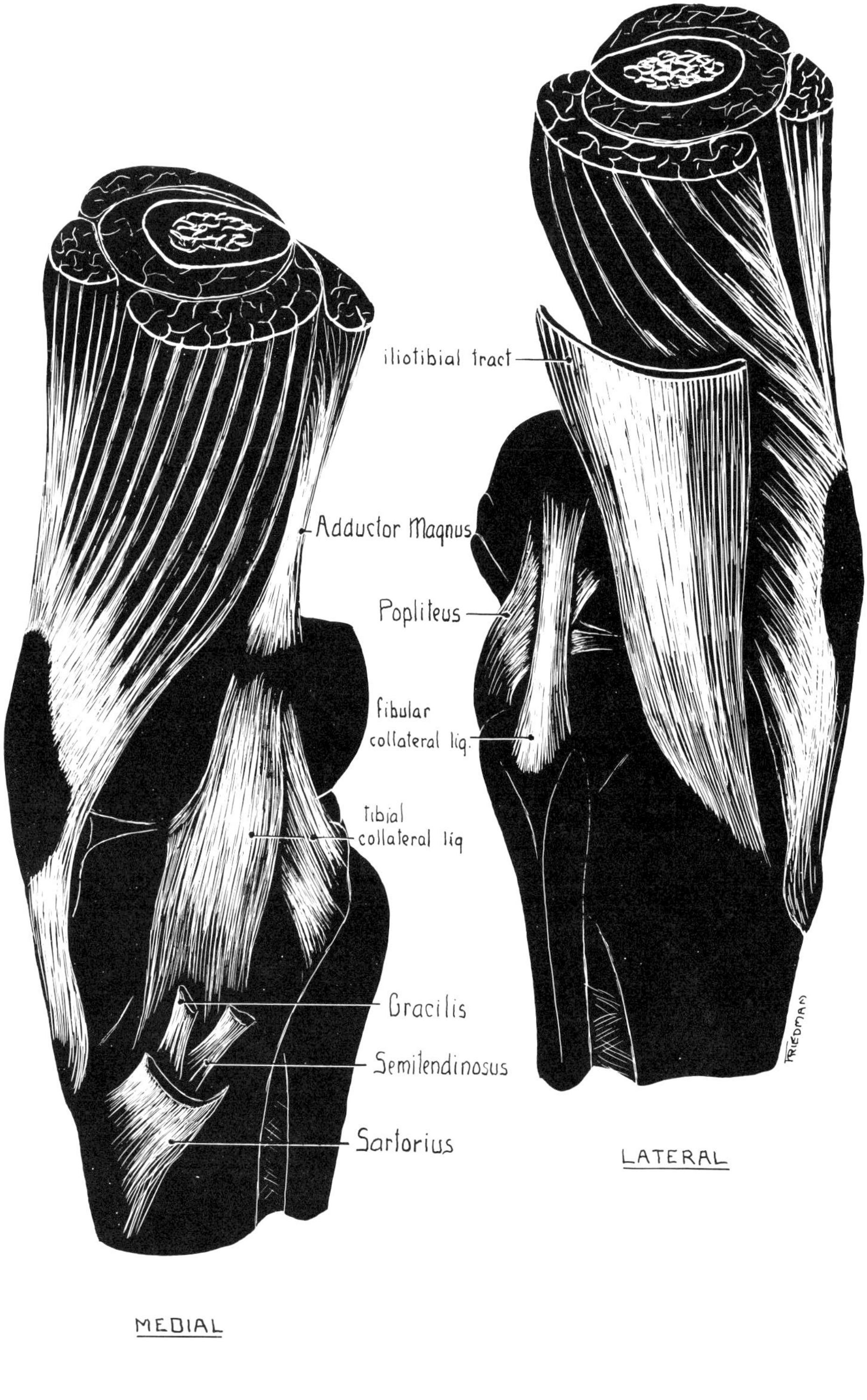
iliotibial tract
Adductor Magnus
Popliteus
fibular
collateral lig.
tibial
collateral lig
Gracilis
Semitendinosus
Sartorius
LATERAL
FRIEDMAN
MEDIAL

The Posterior Stabilizing Ligaments

In general, strong ligamentous bands pass upward from the tibia, and to a lesser degree from the fibula, to the back of the femur. These ligaments pass across the synovial membrane at the back of the femoral condyles and insert just beyond the articular margins. They also bridge the intercondylar notch of the femur so that the cruciate ligaments, although extrasynovial as already described, are fully intracapsular. A constant posterior band, the *oblique popliteal ligament,* extends upward and laterally across this intercondylar region. In addition, ligamentous fibers bridge the tendon of insertion of Popliteus. It is worth noting that the medial, lateral, and posterior stabilizing ligaments as well as the anterior cruciate ligament are all tightest in the fully locked position of the knee, i.e., after medial rotation of the femur.

Two muscles closely related to this portion of the capsule reinforce it. **Popliteus** arises from the back of the tibia between the condylar plateau and the soleal line. The latter is an oblique ridge which extends diagonally downward and medially across the upper third of the back of the tibia. Popliteus inserts by flat tendinous fibers into the back and side of the lateral meniscus of the knee joint and by a denser, more cordlike portion into the lateral epicondyle of the femur. **Semimembranosus** is the deepest of the medially placed hamstring group of muscles at the back of the thigh. Its lower half is a flat dense tendon which passes directly across the back of the knee joint to insert into the medial tibial condyle, into the area of bone below this, and in addition into the fascia overlying Popliteus.

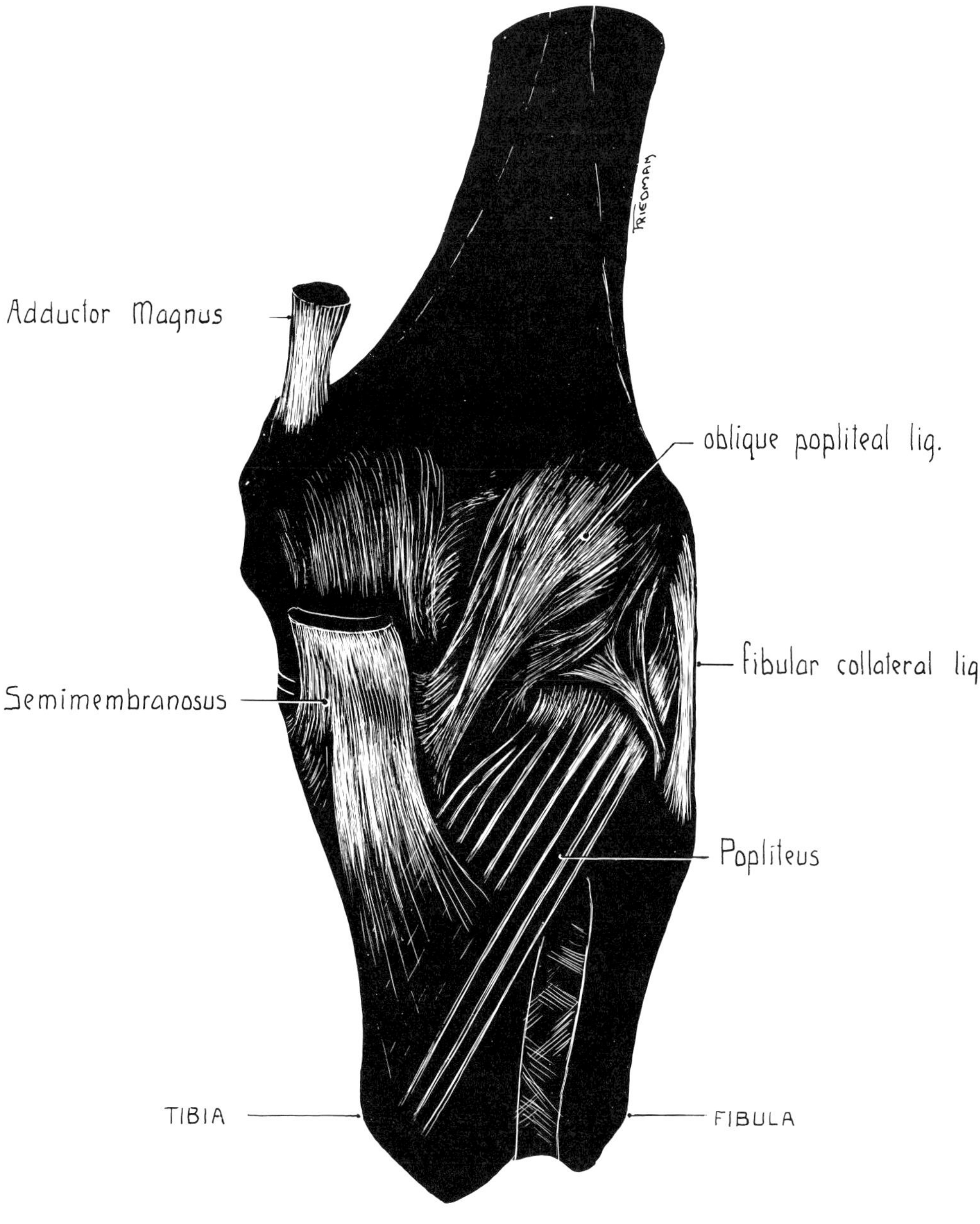
Adductor Magnus
oblique popliteal lig.
fibular collateral lig.
Semimembranosus
Popliteus
TIBIA
FIBULA
FRIEDMAN

POSTERIOR

The Back of the Thigh

The Deep Muscular Plane

Adductor Magnus has already been described as a plane of muscle limiting the front of the thigh. Here it is seen from its posterior aspect as the limiting deep structure of the back of the thigh. Above it lies Quadratus Femoris in the gluteal region.

Adductor Magnus arises fanlike from the inferior ramus of the pubis and ischium as far back as the ischial tuberosity. Its ischial fibers of origin are innervated separately by the sciatic nerve and form a recognizable vertical band that inserts by a round tendon on the adductor tubercle. The remaining fibers, innervated by the obturator nerve, form a flat thick sheet that inserts into a continuous line along the lateral lip of the linea aspera from Quadratus Femoris above to the lower third of the femur below. The lower fibers reach the upper part of the medial supracondylar line.

It is convenient at this time to describe the origin of the upper calf muscles, since these are closely related to the back of the knee.

Gastrocnemius arises by two heads. The medial head takes origin from the back of the medial condyle of the femur just above the capsule. The lateral head arises similarly from above the lateral condyle but extends down over its lateral aspect as well. Both heads pass downward to form much of the bulk of the calf. The horseshoe-shaped origin of **Soleus** from the soleal line of the tibia and from the back of the upper quarter of the fibula is seen here as well.

Plantaris arises immediately above the lateral head of Gastrocnemius. Like Palmaris Longus, it consists of a very short muscle belly and a very long tendon. The tendon passes between the two heads of Gastrocnemius and Soleus. This will be described more fully with the back of the leg (Fig. 74).

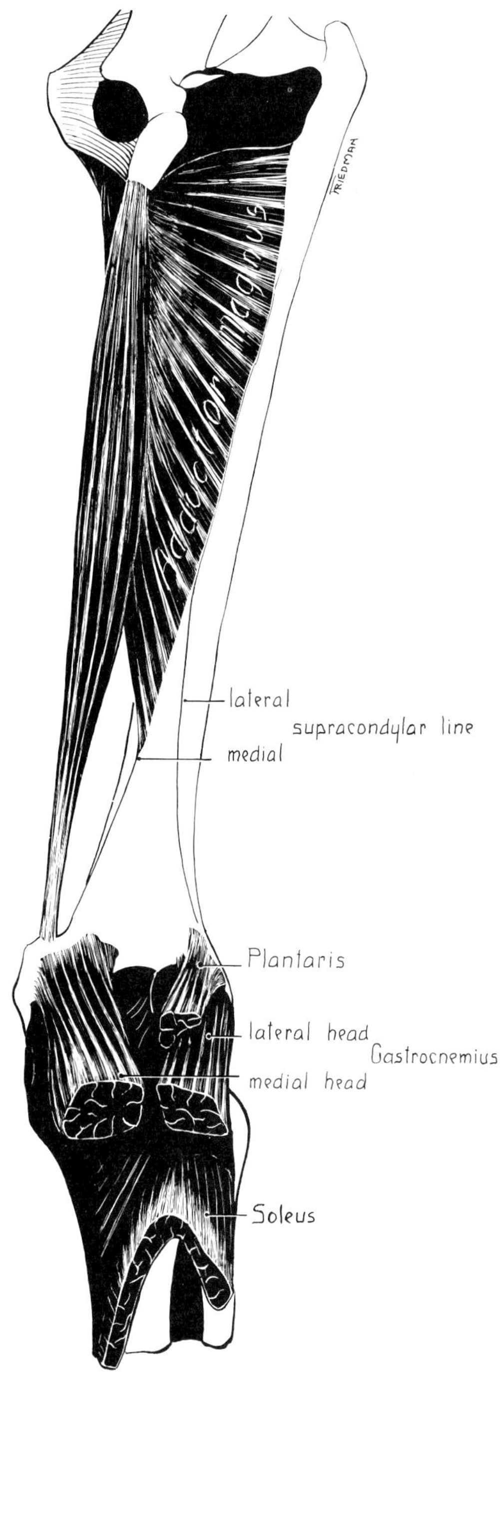
Friedman
Adductor magnus
lateral
supracondylar line
medial
Plantaris
lateral head
Gastrocnemius
medial head
Soleus

The Popliteal Artery and Vein

The femoral artery with its vein behind was seen to leave the thigh through the opening between the two portions of Adductor Magnus. As they enter the back of the thigh, these vessels are renamed **popliteal artery** and **popliteal vein.** The popliteal artery runs downward across the back of the supracondylar area of the femur and here makes direct contact with bone and joint capsule. It continues down on Popliteus and ends by dividing into the anterior and posterior tibial arteries as it is about to leave this muscle.

The popliteal artery gives numerous branches throughout its course. For the most part, these run to supply adjacent muscles, the femur, the tibia, and skin. The vascular anastomosis around the knee joint, like that around the elbow, is exceedingly rich. Five of the branches participating in this are sufficiently constant to warrant separate mention. They are the *lateral* and *medial superior genicular arteries* just above the joint, the *lateral* and *medial inferior genicular arteries* below the joint, and the *middle genicular artery* which passes as a single vessel in the midline into the back of the joint.

A tendinous arch which bridges the tibial and fibular origins of the Soleus muscle provides a tunnel for passage of both tibial arteries into the posterior compartment of the leg. The *posterior tibial artery,* the larger of the two terminal branches, continues down deep to Soleus for a short distance before giving off the *peroneal (fibular) artery*. The *anterior tibial artery* at once passes forward to the anterior compartment of the leg, running between the tibia and fibula just above the interosseous membrane which unites them.

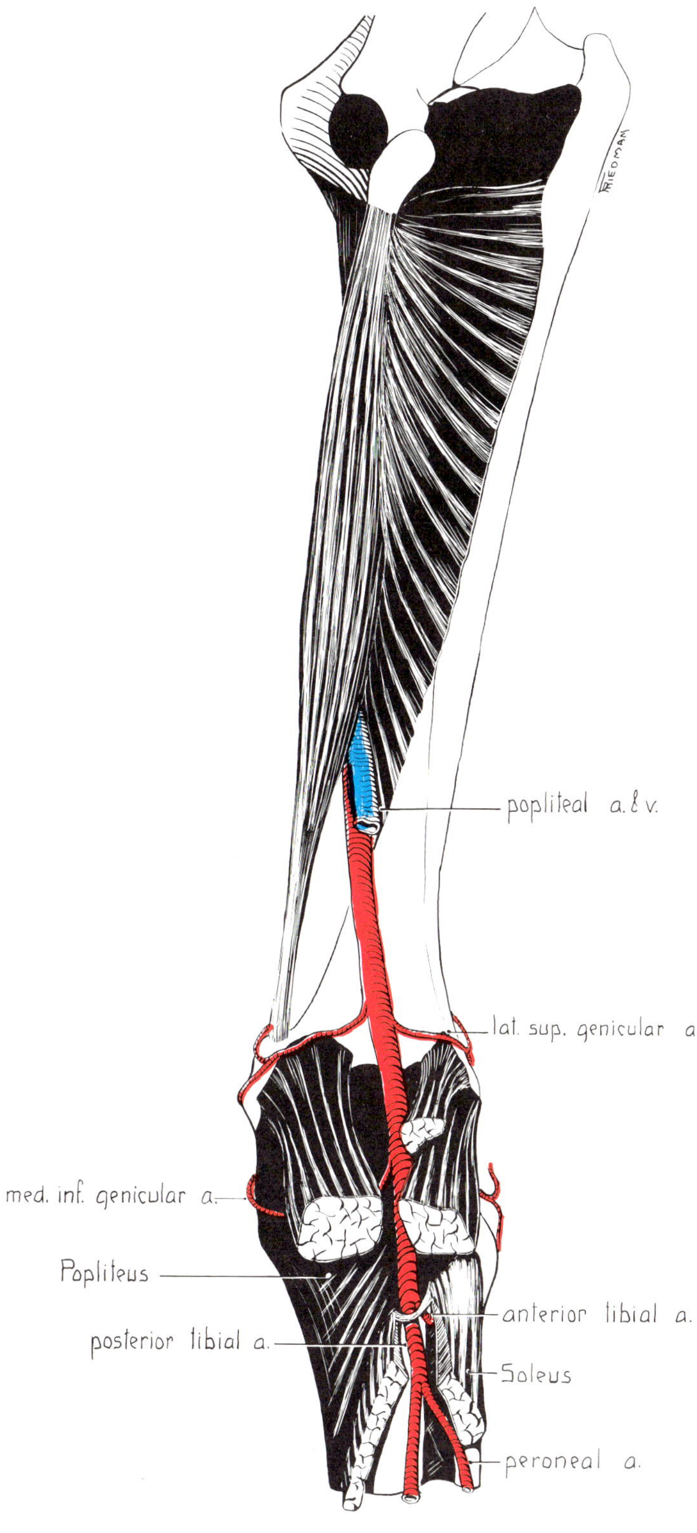
FRIEDMAN
popliteal a. & v.
lat. sup. genicular a.
med. inf. genicular a.
Popliteus
anterior tibial a.
posterior tibial a.
Soleus
peroneal a.

The Sciatic Nerve

It is convenient at this time to summarize some of the general principles underlying muscular attachments at the back of the thigh.

The linea aspera provides attachment for the Adductor group of muscles; of these, Adductor Magnus, the deepest in position, of necessity attaches nearest to the lateral lip of the linea. Here it is almost continuous with Vastus Lateralis, separated from it only by the linear origin of the short head of Biceps Femoris below and the rough gluteal tuberosity for the insertion of Gluteus Maximus above.

Inferiorly, the linea aspera widens out to form the broad, triangular, popliteal surface of the femur which remains free of muscle attachments. The lateral and medial lips of the linea aspera become respectively the lateral and medial supracondylar lines. The short head of Biceps Femoris arises from the lateral line, while a small portion of Adductor Magnus extends for a short distance along the medial line. As in the front of the thigh, Gracilis forms the medial boundary for the thigh muscles, and its long tendon of insertion passes down over the medial portion of the capsule of the knee joint to the tibia, where it inserts under cover of the tendon of Sartorius (Fig. 63).

The course of the **popliteal artery** has been described. It extends from the aperture in Adductor Magnus above to the aperture in Soleus below. In its course, it lies superficial to the popliteal surface of the femur, the back of the capsule, and the back of the Popliteus muscle. As the artery passes through Adductor Magnus, its companion, the **popliteal vein,** lies behind it and remains in this position throughout its course. It is thus always superficial to the artery when exposed from behind. It almost covers the artery superiorly but tends to move somewhat medial to it at the back of the knee.

The **sciatic nerve** runs with a gentle curve through the gluteal region. It crosses Obturator Internus with the associated Gemelli, then Quadratus Femoris, and continues down on Adductor Magnus close to the midline of the back of the thigh. While still on Adductor Magnus it usually divides into tibial and common peroneal nerves. The *tibial nerve* is the larger and continues the direct course of the sciatic. It thus crosses superficial to the popliteal vessels and moves somewhat medial to them at the back of the knee. It should be emphasized that while tibial nerve, popliteal vein, and popliteal artery lie in that order from medial to lateral sides, they are also, and more importantly, in that same order from the skin inward.

The *common peroneal nerve* is directed downward and laterally so as to spiral around the knee to reach the front of the leg. It is at first restrained in position against the back of the femur by the origin of the short head of Biceps Femoris. Emerging below this, it crosses Plantaris and the lateral head of Gastrocnemius to reach the lateral side of the neck of the fibula.

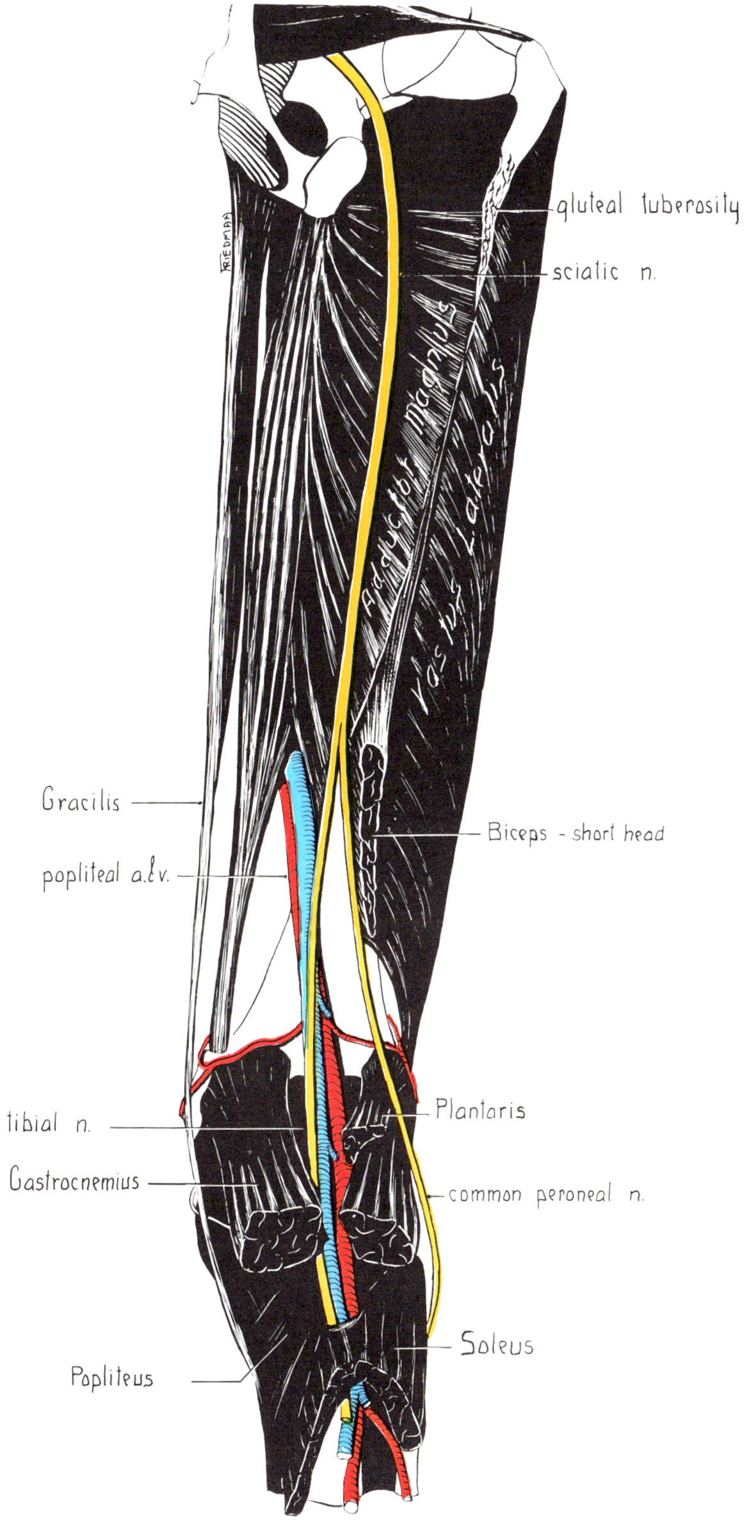
gluteal tuberosity
sciatic n.
Adductor magnus
Vastus Lateralis
Gracilis
Biceps - short head
popliteal a.&v.
Plantaris
tibial n.
Gastrocnemius
common peroneal n.
Soleus
Popliteus

The Superficial Muscular Plane

The origin of Gastrocnemius and Plantaris is from the supracondylar region of the femur. The two heads of Gastrocnemius converge and obscure the tibial nerve and the popliteal vessels. Plantaris lies at first above the lateral head of the Gastrocnemius but quickly becomes tendinous and slides deep to Gastrocnemius to lie between it and Soleus.

The superficial muscular plane of the back of the thigh is formed only by the hamstring muscles—Semitendinosus, Semimembranosus, and Biceps Femoris. These originate narrowly from the back of the ischial tuberosity superiorly and become progressively broader. Toward the back of the knee joint, Biceps separates from Semimembranosus and Semitendinosus and so defines the popliteal fossa.

Semitendinosus consists of a round muscle belly in its upper half and a round tapered tendon in its lower half. Throughout its course it overlies Semimembranosus and toward its insertion passes alongside the tendon of Gracilis to insert with it on the medial aspect of the tibia. **Semimembranosus** consists of a flat membranous sheet in its upper half and a fat muscle belly in its lower half. It becomes flat and tendinous again toward its area of insertion, just below the medial condyle of the tibia, which it reaches by passing deep to the medial head of Gastrocnemius. **Biceps Femoris** consists of a stout fleshy belly throughout most of its length. It receives additional fibers into its deep surface from a short head which arises inferiorly from the linea aspera and the lateral supracondylar line. The final tendon of insertion passes around the lateral side of the knee into the top of the head of the fibula, embracing the insertion of the fibular collateral ligament.

The **popliteal fossa** is a diamond-shaped interval bounded above by Biceps on its lateral side and by Semimembranosus with Semitendinosus medially. Inferiorly, it is bounded by the lateral head of Gastrocnemius together with Plantaris on one side and the medial head of Gastrocnemius on the other. It contains the *tibial nerve, popliteal vein,* and *popliteal artery* in that order from superficial to deep, more or less in the midline, while the *common peroneal nerve* follows its lateral margin. In addition to these primary structures, the fossa contains the *genicular arteries* and *veins, lymph nodes* interrupting the lymphatic drainage of the leg and foot, the termination of the *short saphenous vein,* and *cutaneous nerves* which supply the back of the calf and continue through the *sural nerve* to the lateral side of the ankle and foot as far forward as the little toe.

The **tibial nerve,** while still ostensibly a part of the sciatic nerve, gives branches to all of the hamstring muscles, i.e., to long head of Biceps, to Semimembranosus, and to Semitendinosus, as well as to the medial portion of Adductor Magnus. In the popliteal fossa, the tibial nerve supplies Gastrocnemius, Soleus, Plantaris, and Popliteus. It also provides much of the innervation of the knee joint. The **common peroneal nerve** usually innervates only the short head of Biceps and contributes to the cutaneous innervation of the back of the knee and calf from within the popliteal fossa.

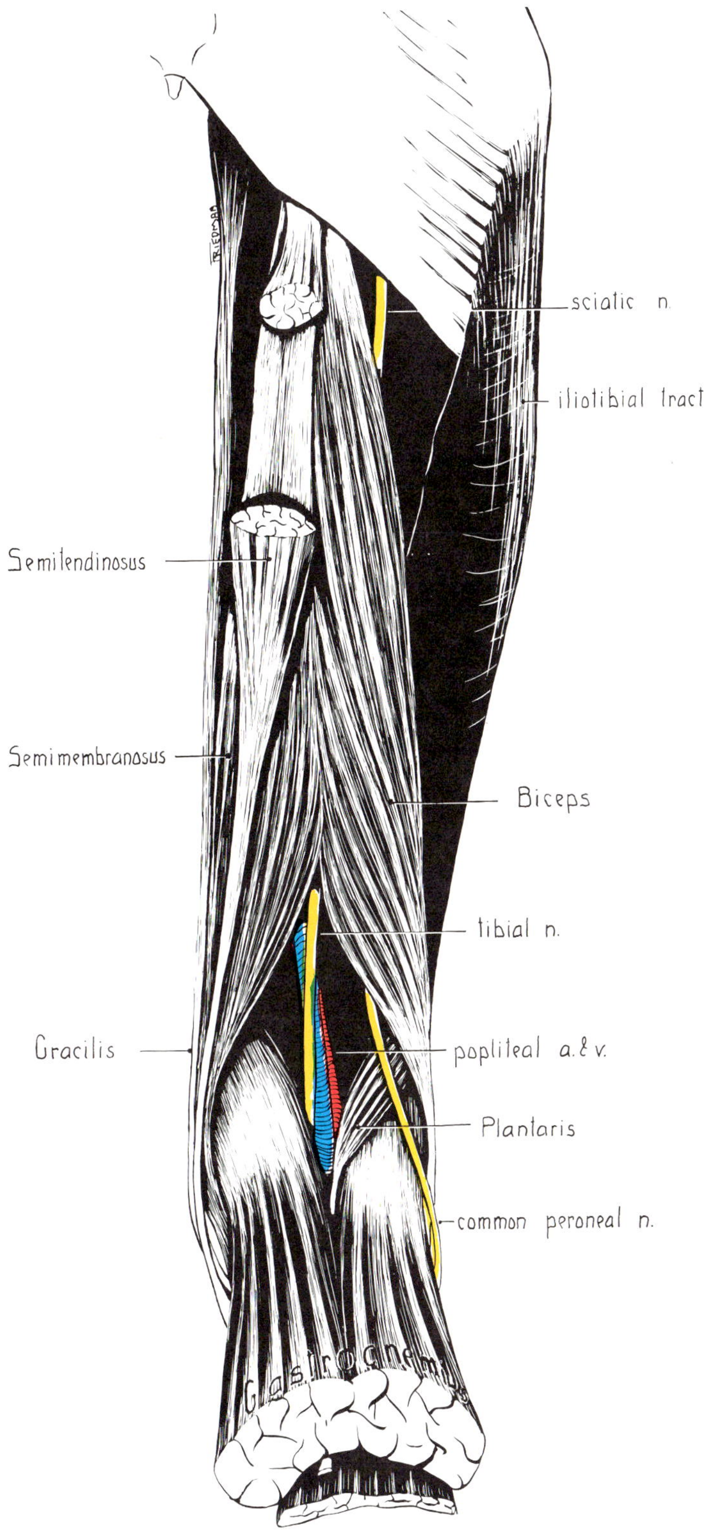
sciatic n.
iliotibial tract
Semitendinosus
Semimembranosus
Biceps
tibial n.
Gracilis
popliteal a. & v.
Plantaris
common peroneal n.
Gastrocnemius

The Leg and Foot

The Arches of the Foot

The organization of muscles in the leg is best understood in terms of the functions of the foot. Thus, the muscles of the front of the leg pass across the dorsum of the foot as extensors, while those of the back of the leg enter the sole or plantar aspect of the foot as flexors. The latter, as they pass around behind the ankle, are forced to the medial side of the heel by the backward protrusion of the calcaneus.

The foot is a highly specialized system of bones arranged to provide a stable base sufficiently strong to support the weight of the whole body on a single limb as well as to provide a shock absorber for the large forces developed in walking, running, and jumping.

The talus is firmly united to the calcaneus, on which it sits, by strong interosseous ligaments. So tightly are these bones joined that there is only a little movement in this *subtalar joint.* Together they constitute a unit that articulates with the bones of the front half of the foot, specifically the navicular and cuboid bones, at the *midtarsal joint.*

The arches are formed by the configuration of the bones of the foot. Indeed, as long as these bones are pulled together tightly, one firmly against the other, both the transverse and longitudinal arches are prominent. In essence the function of the muscles and tendons of the sole of the foot is to retain the close apposition of the bones.

Two arches, longitudinal and transverse, are the shock-absorbing springs. The **longitudinal arch** is the more prominent. It is highest on its medial side where the unit of talus and calcaneus meets anteriorly with the navicular, medial cuneiform, and first metatarsal. The central or pivot point passes through the head of the talus where it is supported by the *"spring"* or *plantar calcaneonavicular ligament.* The lateral part of the longitudinal arch, formed by the calcaneus posteriorly and the cuboid and fifth metatarsal anteriorly, is not so high.

The **transverse arch** is readily seen at the heads of the metatarsals, for here only the head of the first and the head of the fifth metatarsal make contact with the ground, while the middle metatarsals do not. This transverse arch extends back from the heads to the bases of the metatarsals.

It will be appreciated that in standing, the normal foot makes a three-point contact with the ground—at calcaneus, at head of fifth metatarsal, and at head of first metatarsal (supported by an underlying pair of sesamoid bones). This permits the shaft of the second metatarsal to be substantially longer than the first, since it will only make contact with the ground in the last stage of elevating the foot. Should the transverse arch fail, however, the head of the second metatarsal becomes weight-bearing, and since it is a thin bone not structurally designed for this, it may not be able to sustain the load (march fracture).

The toes are used to grip the ground during walking, although this natural action is diminished by the smoothness of shoes. The muscles most responsible for the flexing of the foot on the ankle during walking or running also strongly flex the toes.

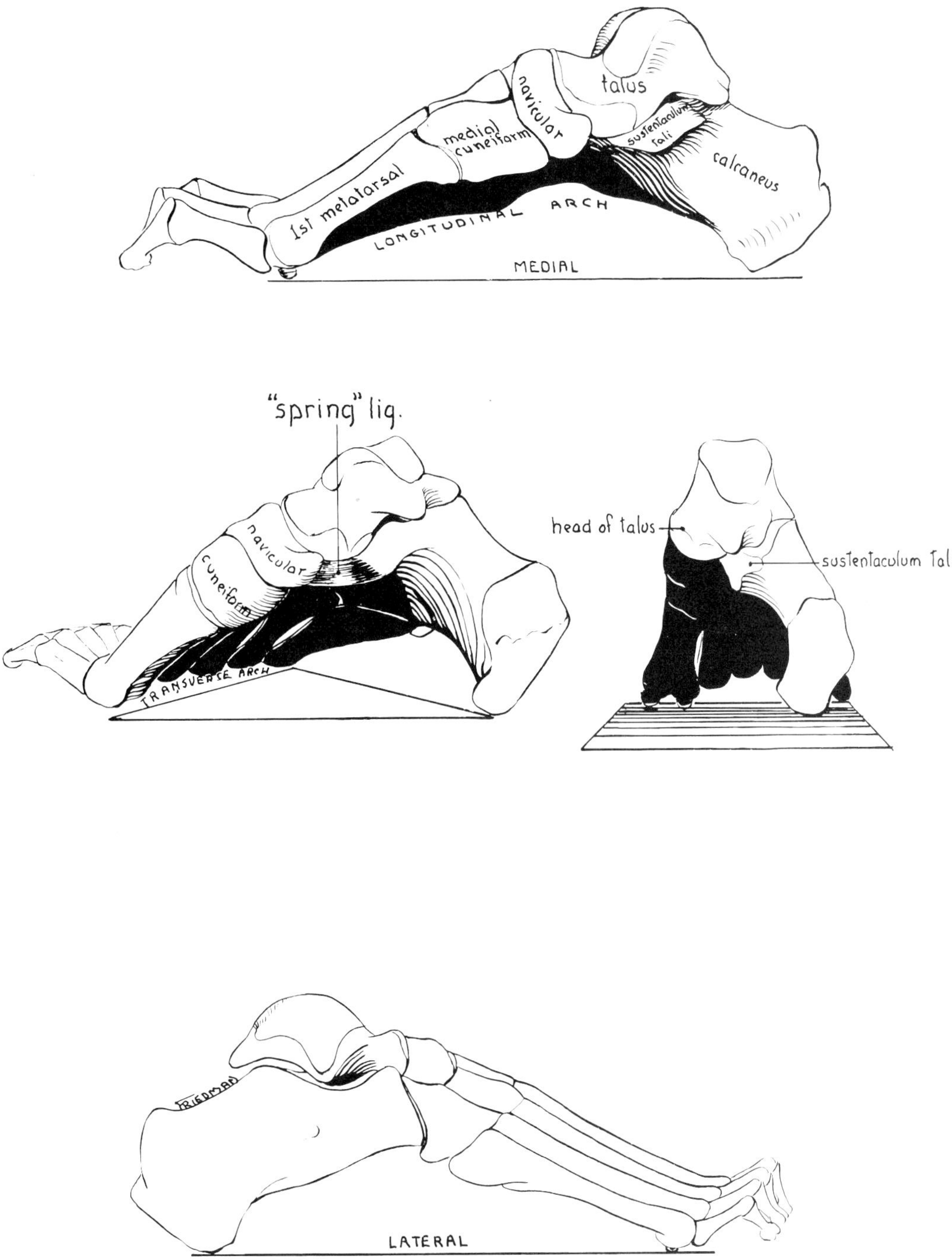
talus
navicular
medial cuneiform
1st metatarsal
sustentaculum tali
calcaneus
LONGITUDINAL ARCH
MEDIAL
"spring" lig.
navicular
cuneiform
TRANSVERSE ARCH
head of talus
sustentaculum tali
FRIEDMAN
LATERAL

The Ankle Joint

The **ankle joint** is the articulation between the tibia and the fibula (united to form a mortice superiorly) and the head of the talus inferiorly. It permits only the simple movement of flexion and extension. The bony flanges provided by the tibial (medial) and the fibular (lateral) malleoli prevent side-to-side slipping. The upper surface of the head of the talus becomes progressively wider anteriorly. As the wider anterior portion moves into the tibiofibular mortice during extension, the bones fit more and more tightly together and the joint acquires added stability.

The *synovial membrane* extends as a simple enclosure joining the articular margins of the bone-pair above to the articular margins of the head of the talus below. This membrane is supported by the usual fibrous capsule thickened to form specific ligaments where this is necessary to provide extra support.

The medial side of the joint is reinforced by one large *deltoid ligament.* This attaches to the malleolus of the tibia above and fans out below to reach the calcaneus along its medial overhang (the sustentaculum tali), then the "spring" ligament, and finally the navicular and the neck of the talus. Not only does this ligament play its part in holding these bones in apposition, but its great size and strength are necessary because it must help to carry the weight of the whole body, which is here cantilevered out over space by the calcaneus.

The lateral side of the joint is reinforced by three separate ligaments. A stout *posterior talofibular ligament,* round and dense, passes between the lateral tubercle on the back of the talus and a deep pit on the inner aspect of the malleolus of the fibula. Without this, the forward thrust of the leg when standing on the toes would be unsupportable. A middle band, the *calcaneofibular ligament,* extends downward from the fibular malleolus to the side of the calcaneus. It is stout but not broad. The third band, the *anterior talofibular ligament,* runs between the neck of the talus and the adjacent part of the fibula. This is the one that is usually injured in a minor "sprained ankle" when the foot has been sharply inverted.

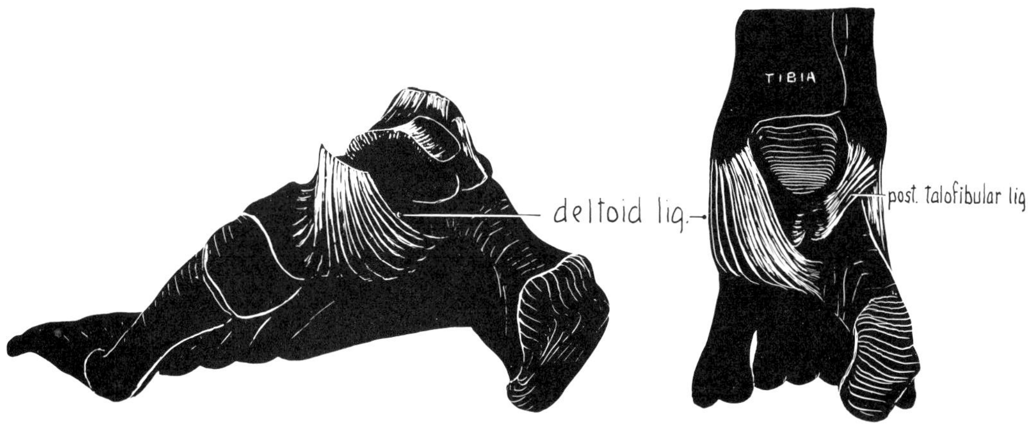
TIBIA
deltoid lig.
post. talofibular lig.

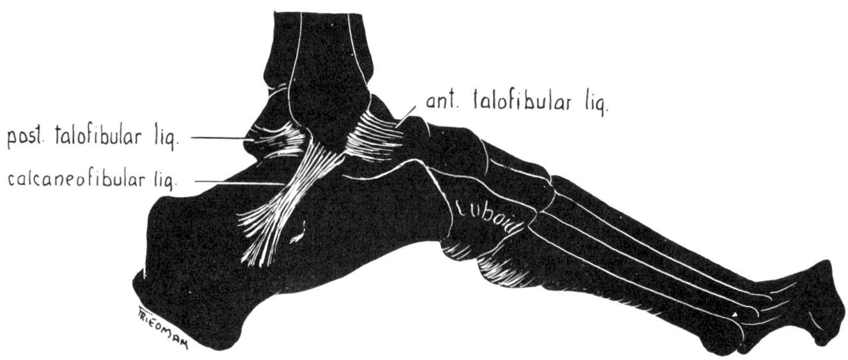
ant. talofibular lig.
post. talofibular lig.
calcaneofibular lig.

The Flexor Compartment:
I. Tibialis Posterior

Tibialis Posterior is the most deeply placed of the flexor group of muscles. It is active in producing flexion of the foot at the ankle joint, but it has little mechanical advantage for this movement. As is usual for muscles that insert close to the joints upon which they act, however, it is admirably placed to support the ligamentous structures of the foot which might otherwise be overstretched when the heel is raised.

Tibialis Posterior arises from the back of the interosseous membrane, which joins fibula and tibia, and from the adjacent portions of both bones below the soleal line. Its tendon, formed in the lower quarter of the leg, passes in a groove behind the tibial malleolus and is directed around the edge of the sustentaculum tali, under the "spring" ligament, and into the tuberosity of the navicular. In addition, tendinous fibers extend the pull of this muscle to all the adjacent bones of the midtarsus.

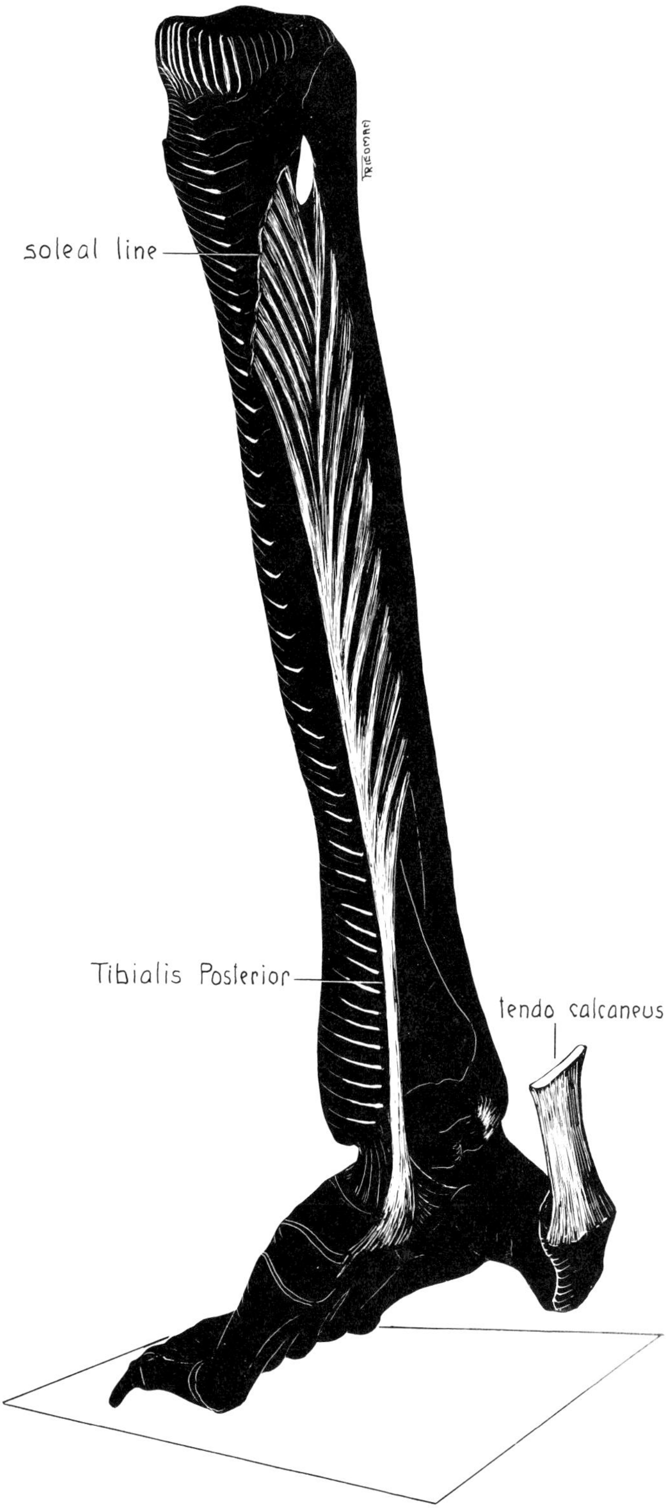
soleal line
Tibialis Posterior
tendo calcaneus

The Flexor Compartment: II. The Long Flexors

The foot is flexed by two groups of structures. One accomplishes the movement by inserting into the heel at the posterior extremity of the foot, the other into the toes at the anterior extremity of the foot. It is apparent that the insertion at the heel carries with it considerable mechanical advantage. If both groups of muscles operate synchronously, the upward pull on the heel is accompanied by a grasping of the ground by the toes and a tightening of the support for the longitudinal arch of the foot.

There are two long flexors both superficial to Tibialis Posterior and originating, in part, from the fascia overlying it. One acts on the big toe, the other on the other four toes. **Flexor Hallucis Longus** arises from the back of the lower half of the fibula below the origin of Soleus. Its tendon is formed opposite the lower end of the tibia and is conveyed through a groove at the back of the talus to the undersurface of the sustentaculum tali, thence across the sole of the foot to the head of the first metatarsal. Here it passes between a pair of sesamoid bones, which lift it from the ground, and across the flat undersurface of the proximal phalanx of the big toe to the base of its distal phalanx.

Flexor Digitorum Longus arises from the back of the tibia below the soleal line and medial to the bony origin of Tibialis Posterior. It thus arises more medially than either Tibialis Posterior or Flexor Hallucis Longus and yet must end in more lateral points of attachment. To accomplish this, its tendon crosses first the Tibialis Posterior tendon and then, running along the back of the tibial malleolus and the side of the sustentaculum tali, enters the sole of the foot and crosses Flexor Hallucis Longus. It divides into four tendons which ultimately insert on the inferior aspect of the distal phalanges of the lateral four toes.

The tendons of Tibialis Posterior, Flexor Hallucis Longus, and Flexor Digitorum Longus are restrained in position against the calcaneus by the *flexor retinaculum.*

Flexor Digitorum Longus is obviously similar to Flexor Digitorum Profundus of the upper limb, just as Flexor Hallucis Longus and Flexor Pollicis Longus are homologous.

Tibialis Posterior
Flexor Digitorum Longus
Flexor Hallucis Longus
Tibialis Posterior

The Flexor Compartment:
III. The Posterior Tibial Artery and the Tibial Nerve

The **posterior tibial artery** is the direct continuation of the popliteal artery and arises from it, together with the anterior tibial artery, at or near the lower border of Popliteus. It then passes into the flexor compartment of the leg bridged over by the tendinous arch in the origin of Soleus. It runs directly down the back of the leg, lying first on Tibialis Posterior and then on Flexor Digitorum Longus under the continuing cover of Soleus. It passes around the back of the ankle between the tendon of Flexor Digitorum Longus anteriorly and Flexor Hallucis Longus posteriorly and then under the sustentaculum tali, and divides into *lateral* and *medial plantar arteries* which enter the sole of the foot. It supplies the muscles in its course and the joints that it passes.

Close to its origin, the posterior tibial artery gives off the large **peroneal artery.** This passes laterally to the fibula and descends into the substance of Flexor Hallucis Longus.

The posterior tibial artery is accompanied throughout its course by the **tibial nerve.** This nerve lies posteromedial to the popliteal artery in the popliteal fossa but crosses the posterior tibial artery to lie lateral to it throughout most of its length. Like the artery, the nerve bifurcates while passing around the sustentaculum tali to form the *lateral* and *medial plantar nerves.*

The **anterior tibial artery** is here seen to pass forward into the anterior compartment of the leg through a gap at the upper border of the interosseous membrane.

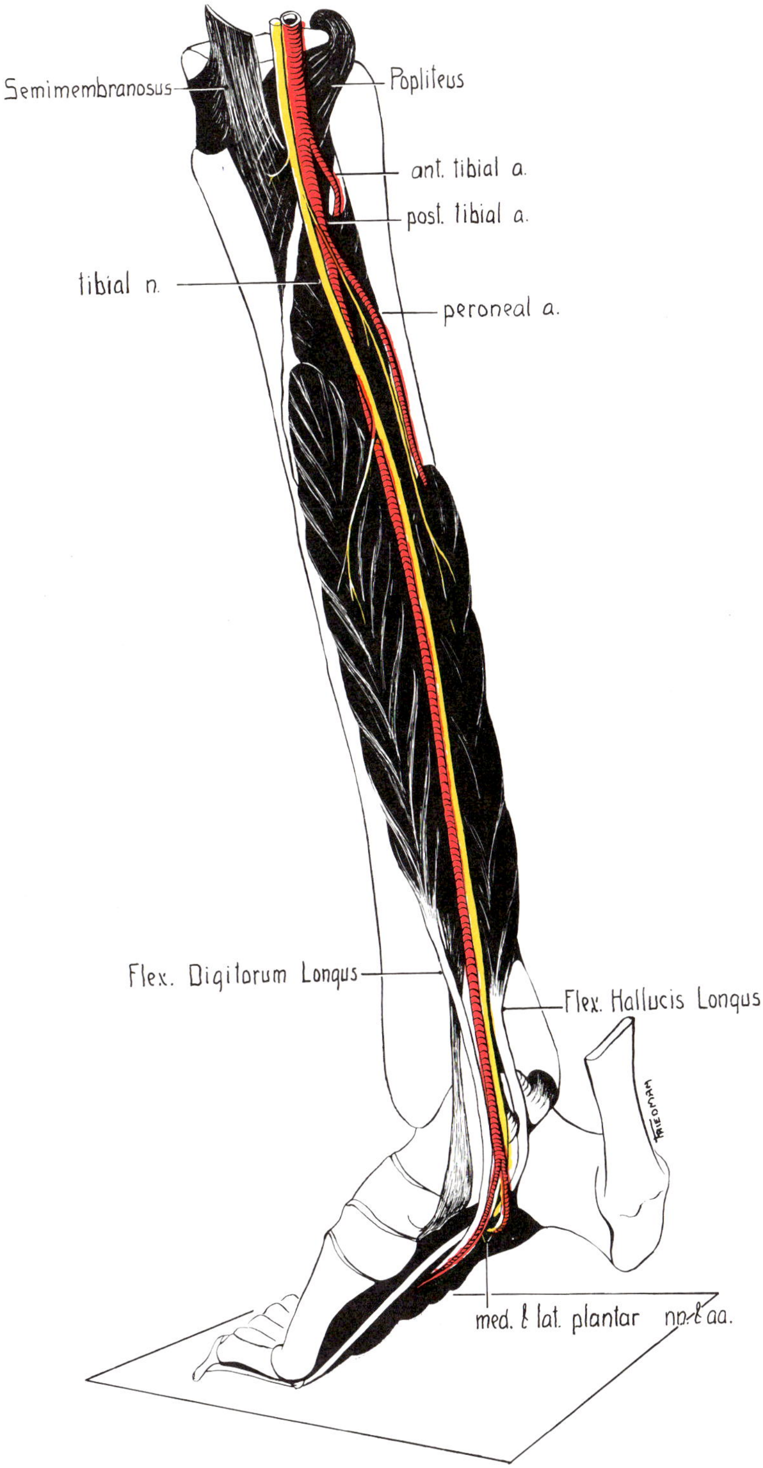
Semimembranosus
Popliteus
ant. tibial a.
post. tibial a.
tibial n.
peroneal a.
Flex. Digitorum Longus
Flex. Hallucis Longus
med. & lat. plantar nn. & aa.

The Flexor Compartment:
IV. The Superficial Flexors of the Ankle

The superficial flexors of the ankle form the large bulky mass of muscle in the calf. **Soleus** was seen to arise from the soleal line on the back of the tibia, from a tendinous arch between the tibia and fibula, and from the posterior aspect of the upper quarter or third of the fibula. From this horseshoe-shaped origin, the fleshy muscle fibers pass downward into an aponeurotic sheet which merges inferiorly with the Gastrocnemii. Soleus covers the deep flexors of the leg throughout most of their length, down almost to the back of the ankle joint. The **lateral** and **medial heads of Gastrocnemius** arise from the respective supracondylar areas of the femur. They form the converging inferior margins of the popliteal fossa and unite in a common tendon which then merges with Soleus to form the one strong, dense *tendo calcaneus* (*Achilles tendon*), which inserts into the posterior part of the calcaneus. A large bursa intervenes between the tendon and the upper part of this bone and is quite prone to inflammation.

Plantaris tendon was earlier seen to arise above the lateral head of Gastrocnemius and then to wind deep to it. The tendon proceeds in the interspace between the Gastrocnemii and Soleus and escapes at the tendo calcaneus. It then runs freely down into the posteromedial portion of the calcaneus.

There is a deep triangular interval formed between the back of the ankle joint, the upper aspect of the calcaneus, and the tendo calcaneus. This interval is occupied only by fat.

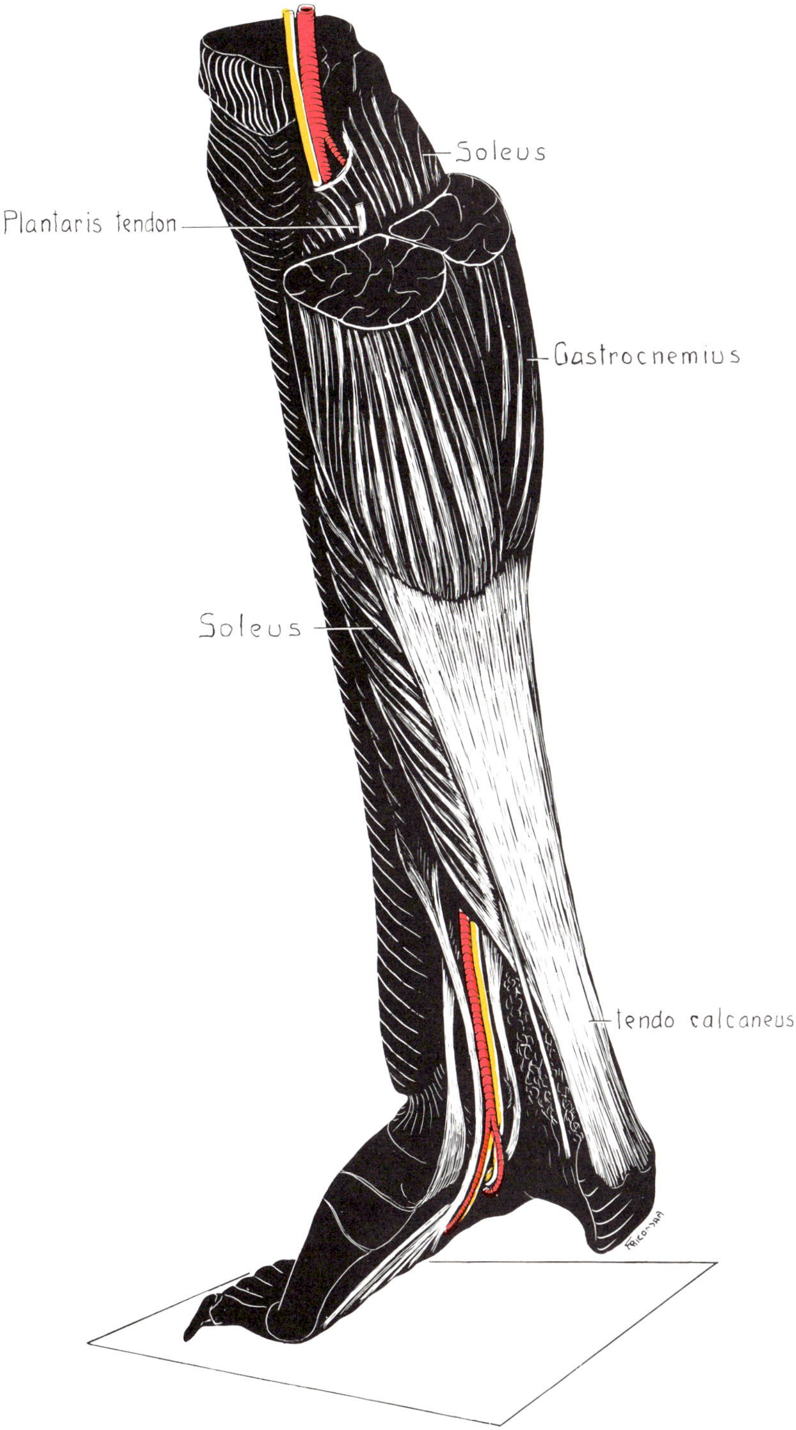
Soleus
Plantaris tendon
Gastrocnemius
Soleus
tendo calcaneus

The Sole of the Foot: The Fourth Plane

The bones of the foot are organized in four functional units. The first and most posterior of these is formed by the **talus** and **calcaneus.**

The second functional unit consists of the five midtarsal bones. The **cuboid** alone forms the entire lateral portion of this unit, while the medial portion is composed of the **navicular** posteriorly and the three **cuneiform** bones (lateral, middle, and medial) anteriorly. This five-bone unit is scooped out inferiorly to form a concavity which continues forward into the metatarsals as the transverse arch of the foot. This second unit meets the talus and calcaneus posteriorly in the midtarsal joint, which runs more or less transversely across the foot. It meets the individual metatarsals anteriorly along a line that runs diagonally forward from lateral to medial.

The third unit of the sole of the foot is formed by the **metatarsal** bones. The first and the fifth are the sturdiest since they bear much of the weight of the body at their heads. The second lies along the long axis of the foot, is the longest, and is firmly mortised into the midtarsal unit so that scarcely any movement is possible here.

Finally, the **phalanges** form the flexible fourth unit. As in the hand, there are three phalanges in each of the four lateral digits and only two in the big toe.

The structures of the sole of the foot are arranged in four layers, by convention numbered from the surface inward. The fourth layer, now to be considered, is thus in the skeletal plane.

The *long plantar ligament* runs from the undersurface of the calcaneus forward across the cuboid to the bases of the lateral metatarsals. Bands from this ligament restrain the tendon of Peroneus Longus. The *short plantar ligament* originates deep to the long but takes a more diagonal course toward the medial tarsal bones. The very thick plantar calcaneonavicular ("spring") ligament forms a bridge between the sustentaculum tali and the adjacent part of the navicular, supporting the longitudinal arch of the foot at its central point. In addition the midtarsal unit of the foot is tightly tied by a series of ligaments which run between the individual tarsal bones and the bases of the metatarsals.

Muscles inhibit overstretching of the ligaments of the midtarsus. **Peroneus Longus** enters the foot by making a sharp bend into a groove on the undersurface of the cuboid. This groove conducts it to its attachments at the adjacent portions of the medial cuneiform and the base of the first metatarsal. It is held in place by dense ligamentous bands which form part of the long plantar ligament. The many tendinous slips of **Tibialis Posterior** radiate widely to complement the support of the midtarsus.

The metatarsals are held firmly to one another by **Plantar** and **Dorsal Interossei.** These are arranged to produce side-to-side movement, with the second digit as axis. It will be appreciated that side-to-side movement of the toes is of considerably less importance than keeping the metatarsals together.

As in the hand, each of the metacarpophalangeal and interphalangeal joints is encased in a dense fibrous capsule. All five are cross-united at the heads of the metatarsals by *deep* and *superficial transverse metatarsal ligaments.*

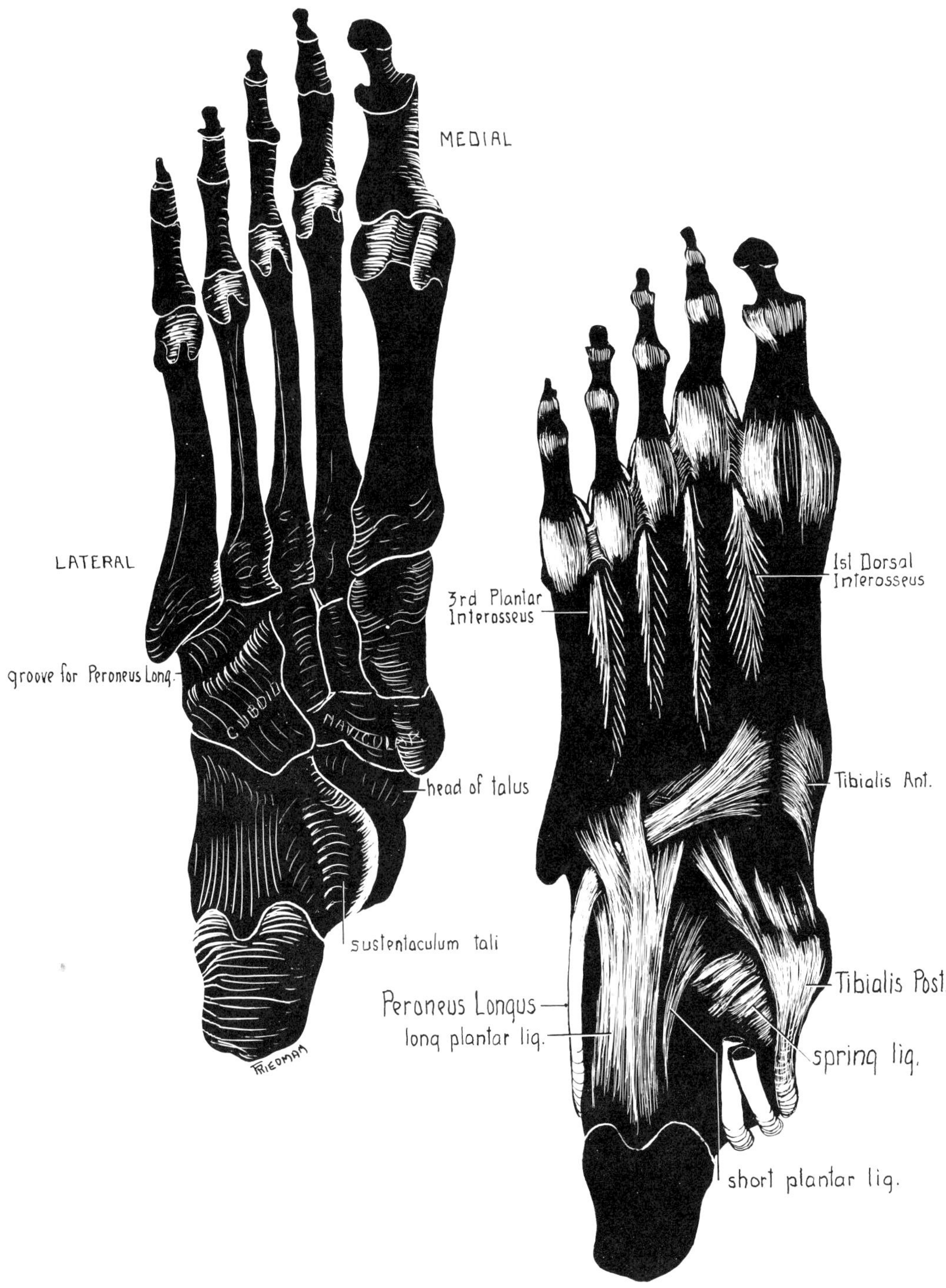
MEDIAL
LATERAL
groove for Peroneus Long.
CUBOID
NAVICULAR
head of talus
sustentaculum tali
FRIEDMAN
3rd Plantar Interosseus
1st Dorsal Interosseus
Tibialis Ant.
Tibialis Post.
Peroneus Longus
long plantar lig.
spring lig.
short plantar lig.

The Sole of the Foot: The Third Plane

Flexor Hallucis Brevis arises in the main from the ligaments overlying the tendon of Peroneus Longus. It passes forward on the undersurface of the first metatarsal and splits into two tendons which insert on the lateral and medial aspects of the first phalanx of the big toe close to the metatarsophalangeal joint. Sesamoid bones are placed in each tendon of insertion, and these raise the tendon of Flexor Hallucis Longus off the ground and also help to keep it in place.

Adductor Hallucis, like its counterpart in the hand, has two heads. The transverse head arises from the ligamentous structures of the metatarsophalangeal joints. The oblique head arises from the midtarsus and from the adjacent tarsometatarsal joints. Both heads insert in common with the medial head of Flexor Hallucis Brevis.

Flexor Digiti Minimi Brevis arises from the lateral ligaments of the tarsus and inserts onto the lateral side of the base of the first phalanx of the fifth toe.

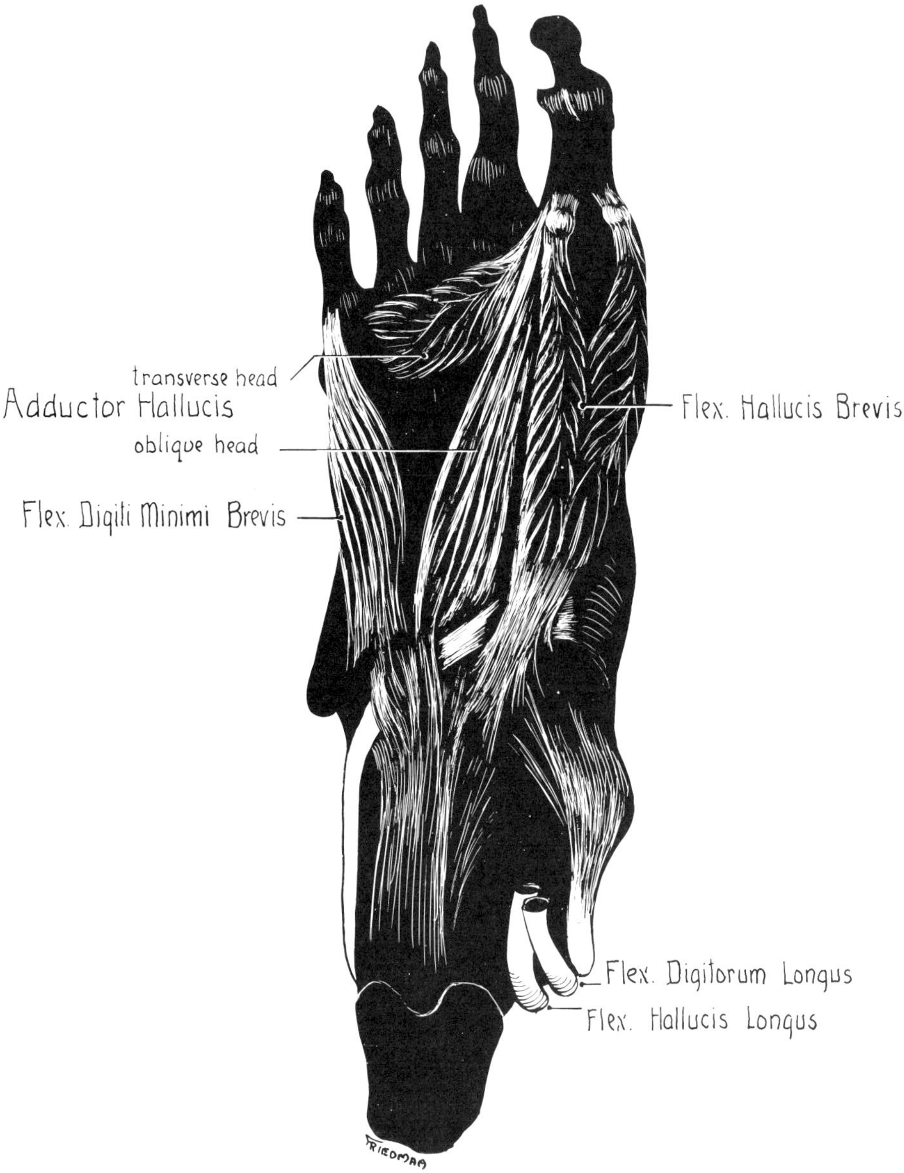
transverse head
Adductor Hallucis
oblique head
Flex. Digiti Minimi Brevis
Flex. Hallucis Brevis
Flex. Digitorum Longus
Flex. Hallucis Longus
FRIEDMAN

The Sole of the Foot: The Second Plane

The long flexors of the toes, Flexor Hallucis Longus and Flexor Digitorum Longus, provide the next layer of support for the longitudinal arch.

The tendon of **Flexor Hallucis Longus** enters the foot as the most lateral of the group of structures that pass beneath the sustentaculum tali. It passes directly forward across the deepest part of the longitudinal arch and enters the big toe between the two sesamoid bones contained in the tendons of Flexor Hallucis Brevis. It inserts into the plantar aspect of the distal phalanx.

Flexor Digitorum Longus enters the foot as a tendon medial to Flexor Hallucis Longus and proceeds to cross the latter superficially on its way to the lateral four toes. Each of the individual tendons enters a flexor sheath at the metatarsophalangeal joint of the toe and is conducted in this tunnel to the distal phalanx. Like its counterpart in the hand, it provides origins for a set of Lumbrical muscles. The first **Lumbrical** arises from the medial side of the first tendon, the remaining three from the adjacent sides of all the tendons. Each Lumbrical passes forward and winds to the back of its digit at the medial side of the metatarsophalangeal joint.

Flexor Accessorius (Quadratus Plantae) is a flat sheet of muscle which arises on the undersurface and medial side of the calcaneus and blends into the lateral side of the tendon of Flexor Digitorum Longus. It tends to realign the otherwise diagonal pull of this long flexor.

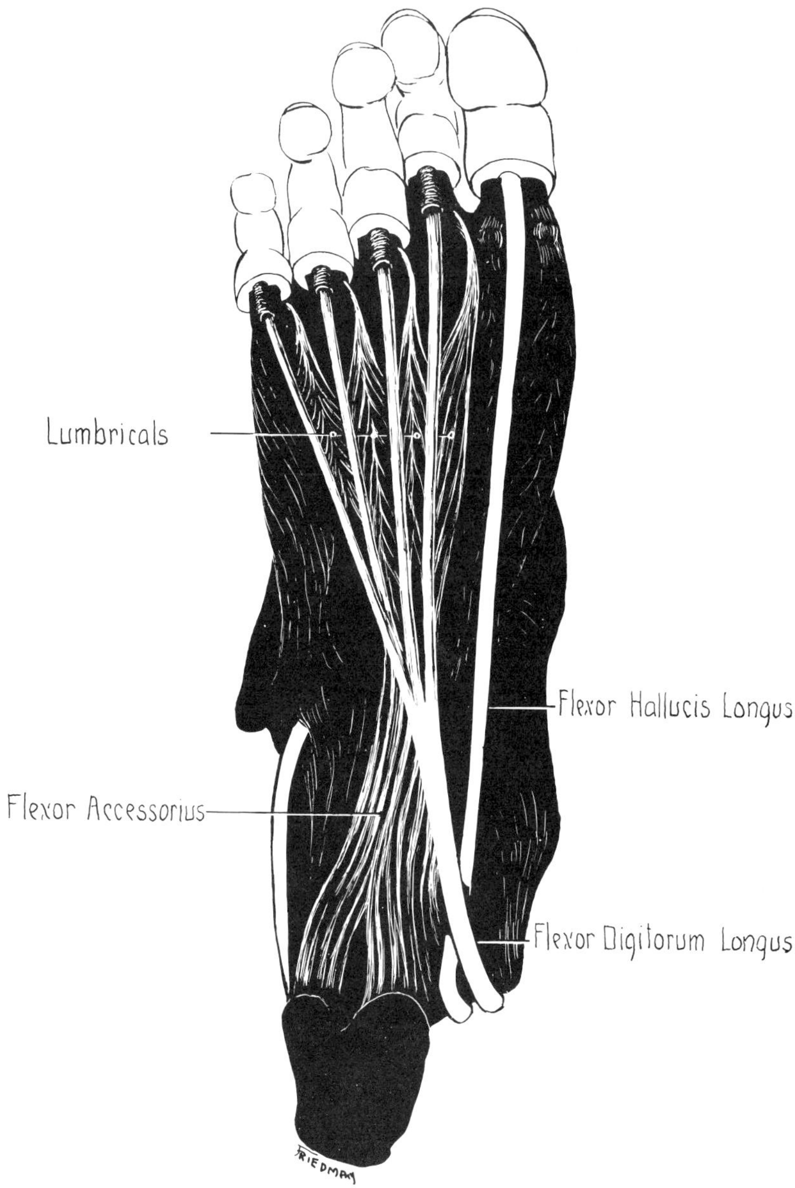
Lumbricals
Flexor Hallucis Longus
Flexor Accessorius
Flexor Digitorum Longus
FRIEDMAN

The Sole of the Foot: The Vessels and Nerves

In general, the vessels and nerves entering the foot lie superficial to the second plane just described. They stay in this position through the posterior half of the foot and then run medially to their final destination between the third and fourth layers.

The **posterior tibial artery** and the **tibial nerve** pass beneath the sustentaculum tali between the tendons of Flexor Hallucis Longus and Flexor Digitorum Longus. Both divide almost at once into lateral and medial plantar arteries and nerves.

The **lateral plantar artery** curves gently laterally toward the front of the foot, then passes deep to the third layer as the *plantar arch*. From this arch, individual *metatarsal arteries* arise which run forward and split at the metatarsophalangeal joint into *digital branches* which supply adjacent sides of each digit. The lateral side of the fifth digit is usually supplied separately. The plantar arch is completed by union with a terminal branch of the *dorsalis pedis artery* (Fig. 80).

The **medial plantar artery** passes forward along the tendon of Flexor Hallucis Longus, supplies both sides of the big toe, and reinforces the digital supply to the second, the third, and usually the fourth toe as well.

The **lateral plantar nerve** distributes itself much as the ulnar nerve in the hand. It thus supplies Flexor Accessorius, at least three of the Lumbricals, the intrinsic muscles of the little toe (Flexor Digiti Minimi Brevis and Abductor Digiti Minimi), Adductor Hallucis, and all of the Interossei. In addition, it provides sensory innervation for all of the fifth digit and some of the fourth.

It remains for the medial plantar nerve to supply, like the median nerve in the hand, the muscles of the "thenar" eminence (Flexor Hallucis Brevis and Abductor Hallucis), the first Lumbrical, and the skin of the plantar aspects of the first three and a half digits or so. As in the hand, the innervation of the plantar aspect of the digits extends to include the dorsal aspect of the terminal phalanx as well.

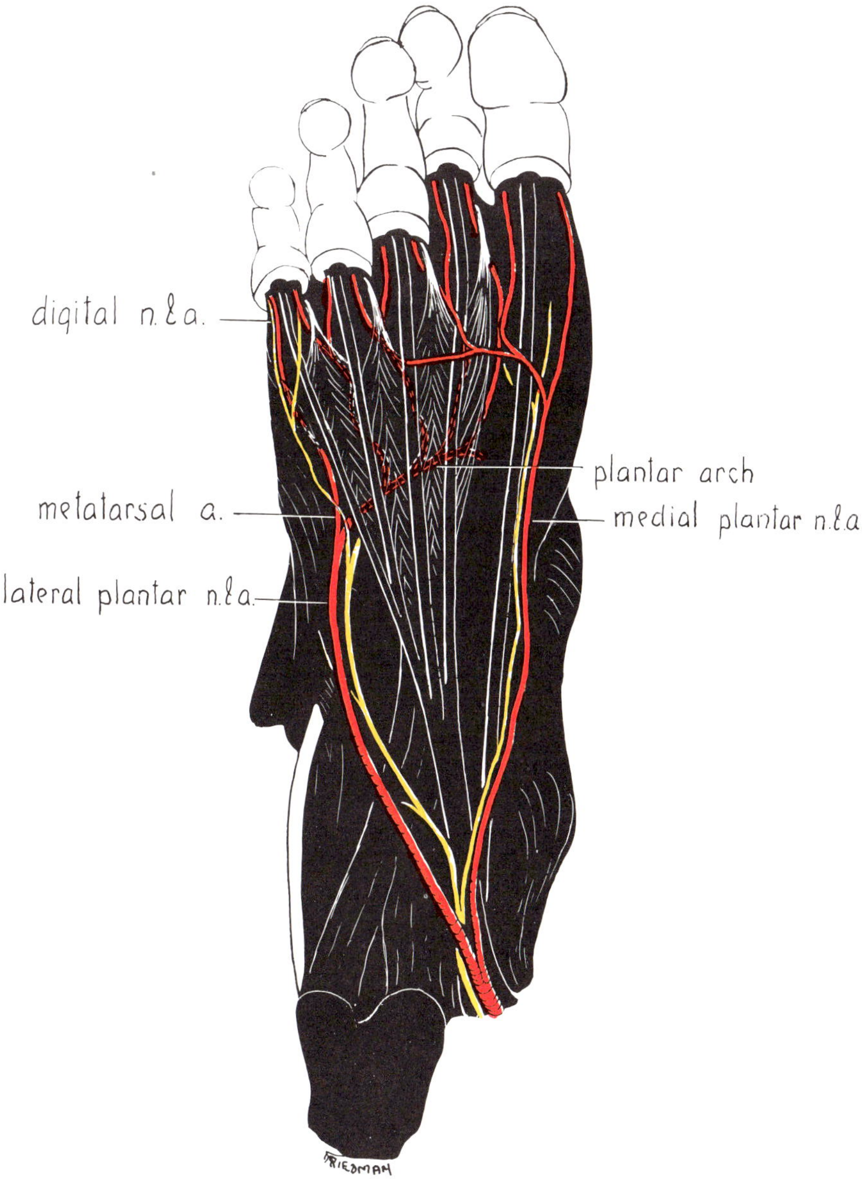
digital n.&a.
plantar arch
metatarsal a.
medial plantar n.&a.
lateral plantar n.&a.
FRIEDMAN

The Sole of the Foot: The First Plane

The first layer of the sole of the foot consists of three large muscles which, running from back to front, support the longitudinal arch and provide much of its spring. Of these, Abductor Digiti Minimi along the shallow lateral portion of the longitudinal arch is the smallest, while Flexor Digitorum Brevis, centrally placed, and Abductor Hallucis Brevis at the deep medial portion of the arch are thicker and fleshier.

Abductor Digiti Minimi differs from its counterpart in the hand in that its origin extends well back of the tarsometatarsal joint right to the tubercles of the calcaneus. It is inserted into the lateral side of the proximal phalanx of the fifth digit. It often has an intermediate insertion into the base of the fifth metatarsal bone.

Flexor Digitorum Brevis does not extend into the calf in the same way as its counterpart (Flexor Digitorum Superficialis) extends into the forearm. It arises from the medial tubercle of the calcaneus and from the dense aponeurotic tissue overlying it. It extends forward and splits into four tendons, each of which enters the fibro-osseous tunnel of the digits accompanied by a tendon of Flexor Digitorum Longus. Within the tunnel, opposite the proximal phalanx, it splits to permit the passage of the long tendon to the terminal phalanx.

Abductor Hallucis arises from the medial tubercle of the calcaneus, from the fascia overlying it, and from the fascia between it and Flexor Digitorum Brevis. It passes forward along the medial side of the first metatarsal to insert with the medial tendon of Flexor Hallucis Brevis.

Overlying the first layer of the sole of the foot is the **plantar aponeurosis.** This consists of a dense central portion encasing Flexor Digitorum Brevis and less dense lateral and medial portions which overlie Abductor Digiti Minimi and Abductor Hallucis respectively. The anterior portion of the plantar aponeurosis splits into columns which follow the line of each metatarsal; these are cross-united at the bases of the digits.

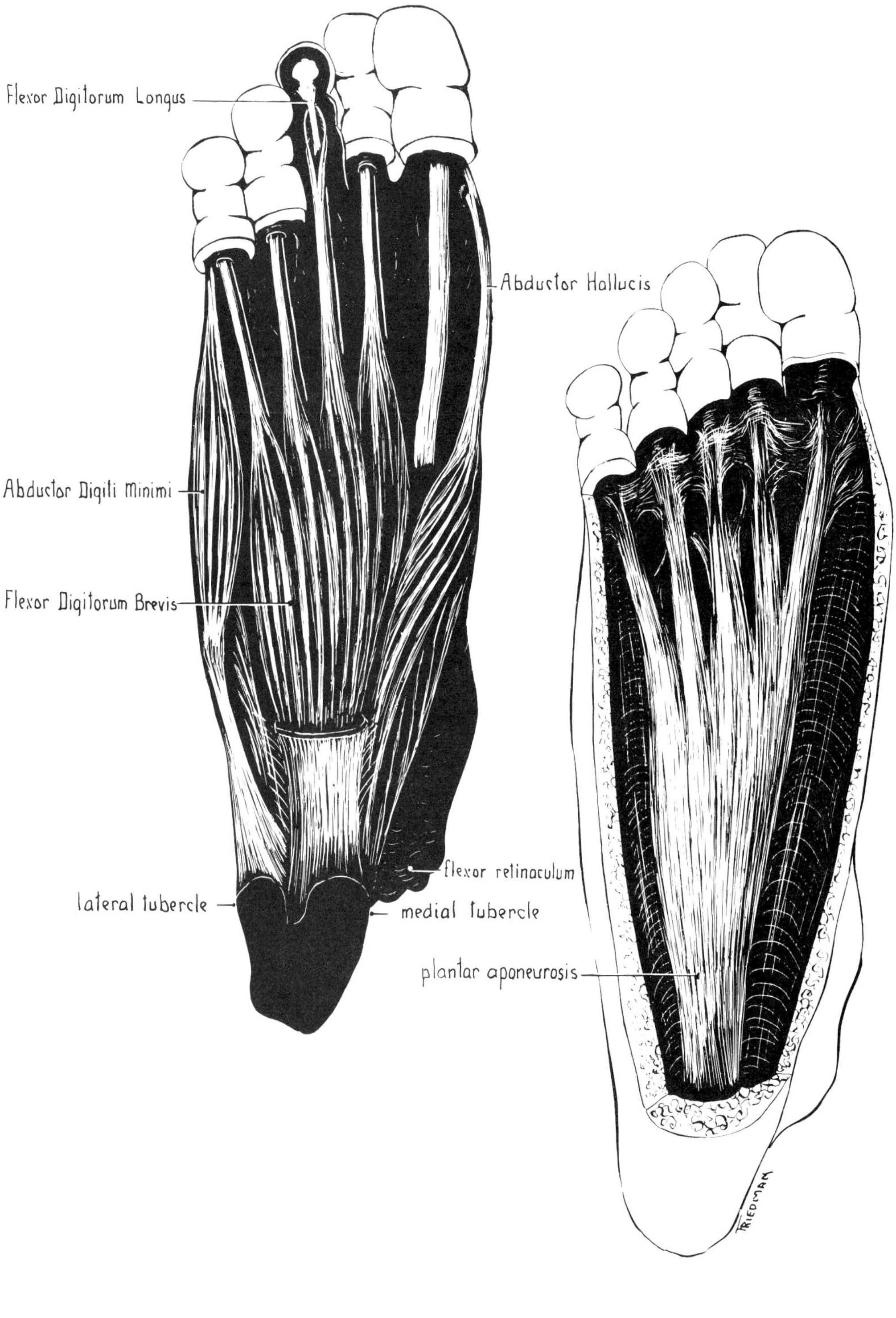
Flexor Digitorum Longus
Abductor Hallucis
Abductor Digiti Minimi
Flexor Digitorum Brevis
flexor retinaculum
lateral tubercle
medial tubercle
plantar aponeurosis
FRIEDMAN

The Peroneal and Extensor Compartments

Only Peroneus Longus and the underlying Peroneus Brevis occupy the peroneal compartment, a fascial enclosure that includes the lateral aspect of the fibula. The lower lateral fibula remains uncovered by any muscle and is a common site for fractures.

Peroneus Longus arises from the upper half of the fibula and, in its course, overlaps the origin of **Peroneus Brevis,** which arises from the lower half of the bone. The tendons of both muscles pass together around the back of the malleolus of the fibula, then along the lateral side of the calcaneus, and separate at the *peroneal tubercle.* Here they are restrained in position by dense fascial sleeves (Fig. 81). Peroneus Longus continues down and forward to reach the lateral margin of the cuboid and is then carried into the groove on the undersurface of this bone and so across the sole of the foot. It is a strong everter of the foot. Peroneus Brevis passes almost directly forward to insert into the dorsal aspect of the tuberosity of the fifth metatarsal bone.

The **common peroneal nerve** was first described in the popliteal fossa, where it follows Biceps Femoris along the lateral side of the knee joint to reach the upper end of the fibula. It then winds around the neck of the fibula to enter the peroneal compartment deep to Peroneus Longus. It splits at once into the *deep peroneal nerve* which passes forward into the anterior compartment and the *superficial peroneal nerve* which continues down with the Peronei. The superficial peroneal nerve innervates both peroneal muscles and ends in cutaneous branches which are distributed over the dorsum of the foot and the lateral digits.

The extensor or anterior fascial compartment extends from the peroneal compartment to the anterior border of the tibia medially. It contains the extensor muscles of the foot accompanied by the anterior tibial artery and deep peroneal nerve.

The **anterior tibial artery,** one of the terminal branches of the popliteal artery, enters the anterior compartment through an opening at the upper border of the interosseous membrane. It runs vertically down the interosseous membrane to reach the anterior surface of the lower end of the tibia. It is here placed midway between the two malleoli and, continuing forward across the tarsus, is renamed the *dorsalis pedis artery.* This vessel passes across the head of the talus, the middle of the navicular, and the middle cuneiform bone to gain the first intermetatarsal interspace. Here it gives off the *arcuate artery,* which arches gently lateralward across the bases of the metatarsals and provides these and their interspaces with branches. The dorsalis pedis artery itself penetrates between the heads of the first Dorsal Interosseus to gain the sole of the foot, and here contributes to the metatarsal and phalangeal arterial supply by anastomosing with the plantar arch.

The **deep peroneal nerve** runs with the anterior tibial artery throughout most of its course. It is at first lateral to the artery but overlaps the vessel in part of its course. It innervates all the extensors of the anterior compartment and ends by supplying the skin bordering the first digital interspace.

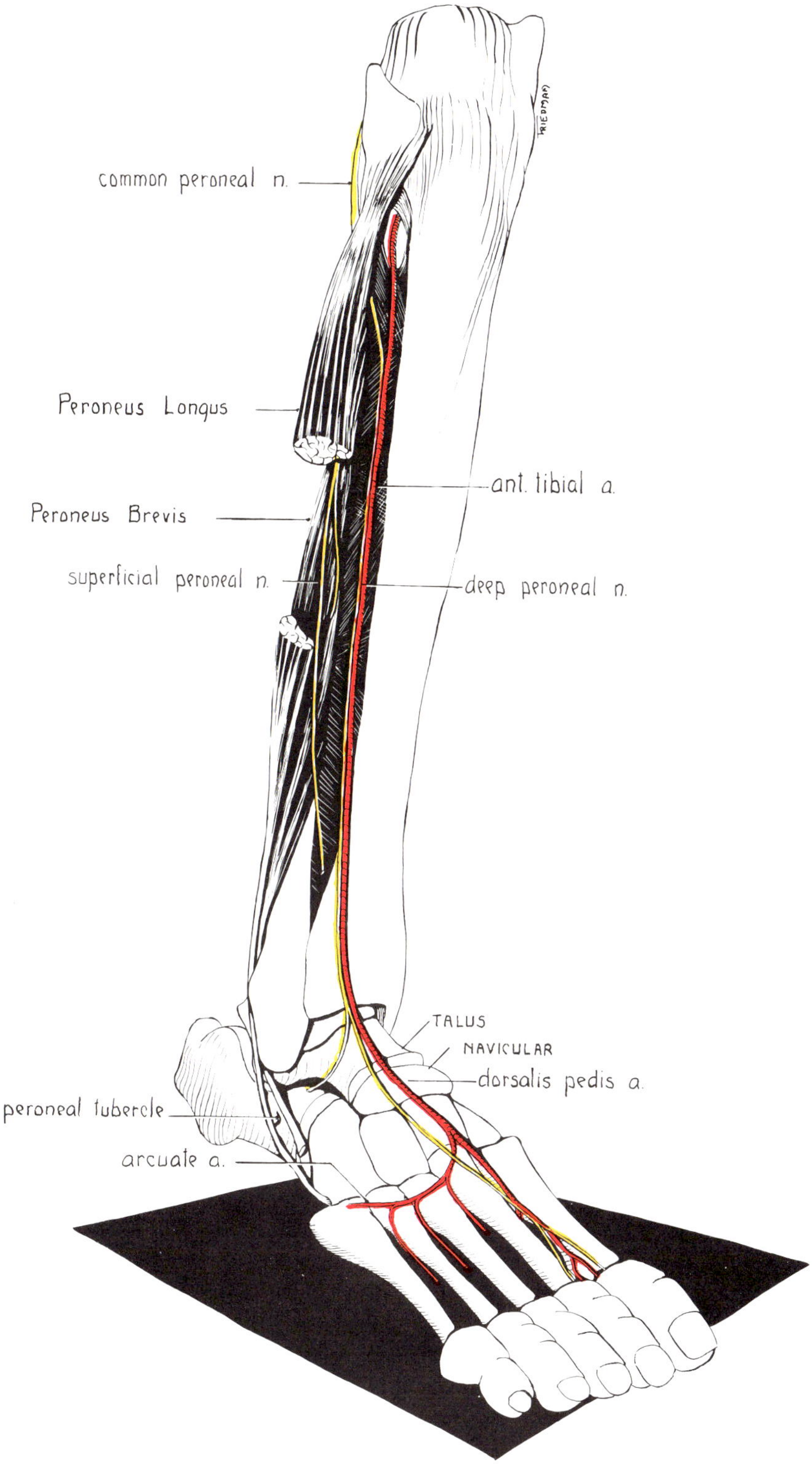
common peroneal n.
Peroneus Longus
ant. tibial a.
Peroneus Brevis
superficial peroneal n.
deep peroneal n.
TALUS
NAVICULAR
dorsalis pedis a.
peroneal tubercle
arcuate a.

The Extensor Muscles:
I. Tibialis Anterior

The extensor muscles of the foot arise from the thin rim of bone which is the anterior surface of the fibula, from the adjacent interosseous membrane, and from the lateral surface of the tibia as far forward as its subcutaneous anterior border.

Tibialis Anterior is the largest muscle of the anterior (extensor) compartment. It originates from all of the lateral surface of the tibia, from the adjacent part of the interosseous membrane, and in a small way from the head of the fibula as well. It tapers to a stout tendon, the most medial of those at the front of the ankle joint, and inserts into the medial side of the first cuneiform and the base of the adjacent first metatarsal bone.

Extensor Digitorum Brevis is considered here although it is not properly one of the muscles of the extensor compartment. It originates from the upper surface of the calcaneus near the midtarsal joint and forms a small belly which overlies the midtarsal bones. It gives rise to four thin tendons which insert into the dorsal expansions on the back of each of the medial four toes. Like the other extensors, it is innervated by the *deep peroneal nerve.* Its most medial tendon crosses the termination of the dorsalis pedis artery.

The extensor retinaculum consists of two portions, an upper transverse band and a lower Y-shaped band. The *superior extensor retinaculum* runs between the anterior borders of the tibia and fibula. It forms a tight restraining band for the extensor muscles and usually provides an individual enclosure for Tibialis Anterior. The *inferior extensor retinaculum* arises from the calcaneus as the continuation of the ligamentous bands restraining the Peronei. It extends across the dorsum of the foot and divides into one band that passes upward toward the lower end of the tibial malleolus and another that passes around to the medial side of the navicular.

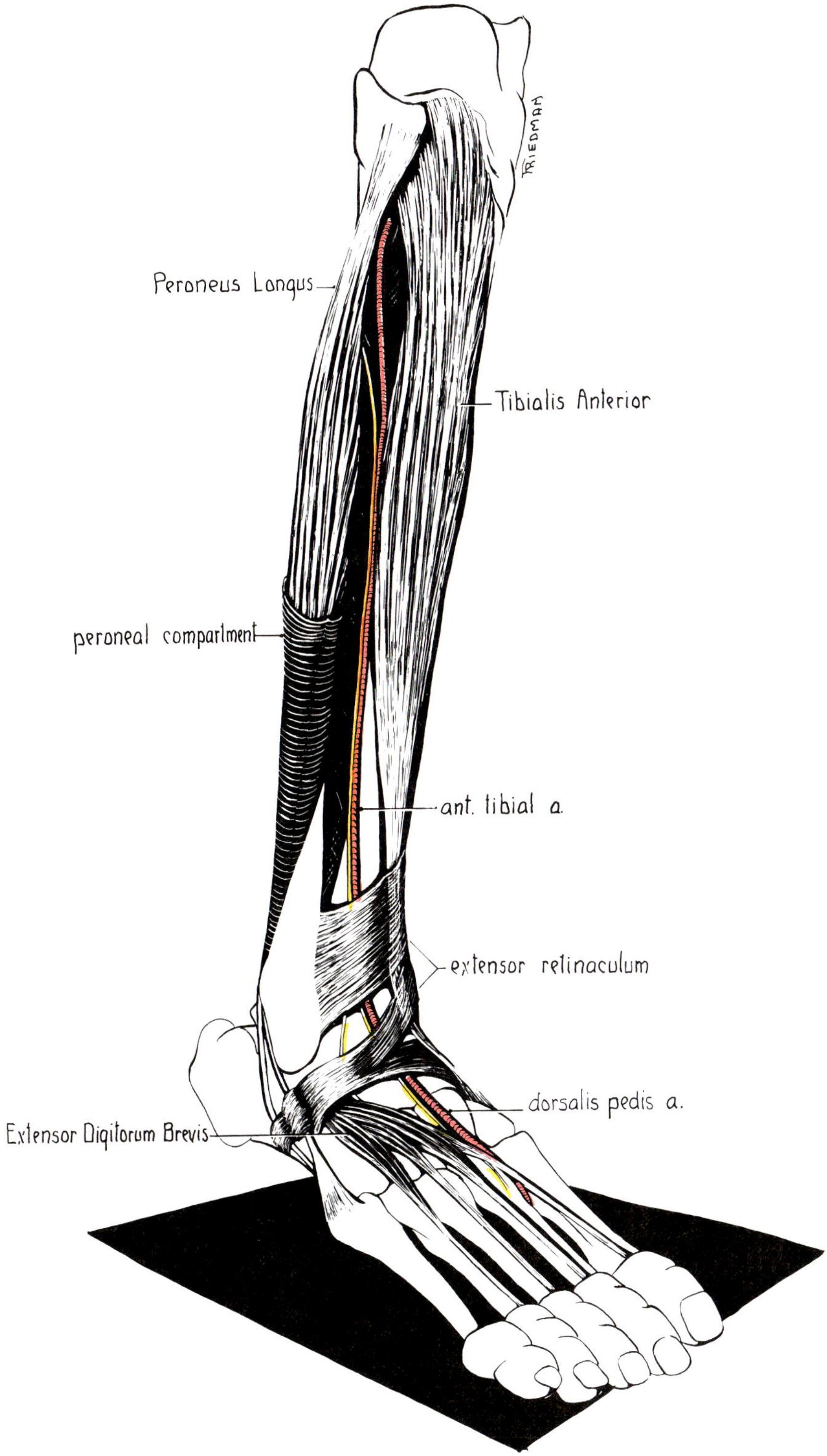
FRIEDMAN
Peroneus Longus
Tibialis Anterior
peroneal compartment
ant. tibial a.
extensor retinaculum
dorsalis pedis a.
Extensor Digitorum Brevis

The Extensor Muscles:
II. The Long Extensors of the Digits

Extensor Hallucis Longus arises from the middle half of the narrow anterior aspect of the fibula and from the adjacent interosseous membrane. It is immediately alongside Tibialis Anterior, with the anterior tibial artery and deep peroneal nerve between. Its tendon passes down deep to the superior and inferior extensor retinacula and crosses the dorsalis pedis artery from lateral to medial side as it does so. It extends forward to the dorsal aspect of the proximal phalanx of the big toe. Thus, while the anterior tibial artery lies medial to Extensor Hallucis Longus in the leg, its continuation, the dorsalis pedis artery, lies lateral to the tendon in the foot.

Extensor Digitorum Longus has a long narrow origin which extends from the head of the fibula and the adjacent part of the tibia and down the whole length of the narrow anterior surface of the fibula. The fibers arising from the lower fourth of this continuous line of origin form a separate muscle, **Peroneus Tertius.** Extensor Digitorum Longus overlaps the anterior tibial artery and nerve throughout but is separated from them by Extensor Hallucis Longus in the midpart of the extensor compartment. It forms four tendons which pass to the dorsal expansions of the lateral four toes, crossing Extensor Digitorum Brevis as they do so. The dorsalis pedis artery thus lies between Extensor Hallucis Longus and the first tendon of Extensor Digitorum Longus. Peroneus Tertius passes as a separate tendon with the most lateral part of Extensor Digitorum Longus and inserts into the dorsal aspect of the fifth metatarsal to a variable extent.

All of the tendons of the anterior compartment are tightly bound down by the retinacula and are thus kept from bowing during extension (dorsiflexion) of the foot. Their separate, tubular, synovial sheaths occupy a variable series of separate compartments within the extensor retinacula.

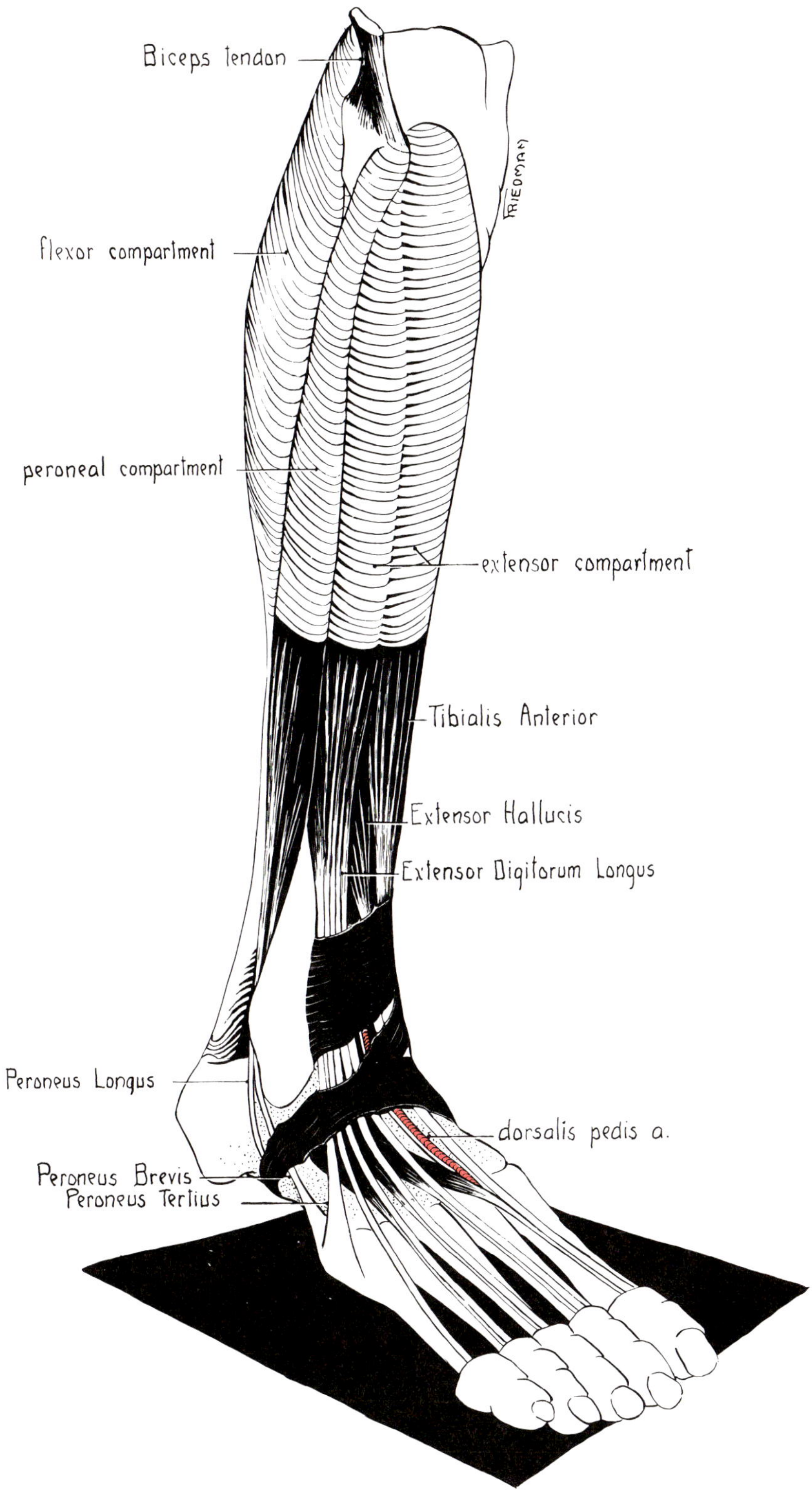
Biceps tendon
flexor compartment
peroneal compartment
extensor compartment
Tibialis Anterior
Extensor Hallucis
Extensor Digitorum Longus
Peroneus Longus
dorsalis pedis a.
Peroneus Brevis
Peroneus Tertius
FRIEDMAN

The Surface Features of the Lower Limb

The Superficial Veins and Lymphatics

The superficial veins of the lower limb require special attention since they are subject to a considerable hydrostatic force throughout life and commonly fail to cope adequately with this. Such failure is marked by increasing tortuosity of the vein in its early stages, then by varicosity followed ultimately by ulceration if untreated. The superficial veins belong to two major systems—the great and the small saphenous veins—which intercommunicate freely.

The **great (long) saphenous vein** begins in the marginal vein of the medial side of the foot. Here it receives blood from the dorsum of the foot as well as some from the sole. Together with its immediate tributaries, the vein passes up in front of the medial malleolus of the tibia and crosses the subcutaneous medial surface of the lower third of this bone obliquely to gain its medial margin. Its crossing of the bone is unsupported and a cardinal site for disease. The vein continues along the medial border of the tibia to the posterior part of the medial condyle of the femur and then up the medial side of the thigh. It winds progressively forward to reach the *saphenous opening* about 3 cm below the inguinal ligament. Here it joins the femoral vein. Throughout its length the great saphenous vein is freely supplied with valves which aid it in supporting the column of blood. It is commonly multiple, especially below the knee (Fig. 59).

The **small (short) saphenous vein** begins with the lateral marginal vein of the foot and passes behind the lateral malleolus. It gradually gains the posterior aspect of the calf and runs up to the popliteal fossa, where it pierces the fascial roof of the fossa to join with the popliteal vein. It too is freely supplied with valves.

The lymphatic drainage of the lower limb follows the plan of the venous drainage. It thus consists of deep channels running with the main arterial distribution and deep veins, and superficial channels lying in the subcutaneous tissue which tend to parallel the course and distribution of the great and small saphenous veins. The most important location for lymph nodes along the course of the lymphatic channels is in the popliteal fossa and in the inguinal region in and about the saphenous opening. The *popliteal lymph nodes* lie mostly within the fat of the popliteal fossa and filter the lymph flow from the deep vessels of the leg and foot as well as from the channels alongside the small saphenous vein. The major mass of inguinal lymph nodes is disposed as superficial and deep inguinal nodes. The *superficial inguinal nodes* form a T, horizontal along the lower border of the inguinal ligament and vertical down to and beyond the saphenous opening. These nodes receive lymph from the lower abdomen, from the external genitalia, and from alongside the great saphenous vein. The *deep inguinal nodes* lie alongside the femoral vein immediately subjacent to the saphenous opening. These nodes interrupt channels that run alongside the drainage system of the femoral vein.

The detailing of cutaneous innervation in the lower limb is of limited practical use. The principal features are illustrated. The dermatomes L4, L5, and S1 in the foot are useful in localizing a herniated intervertebral disc.

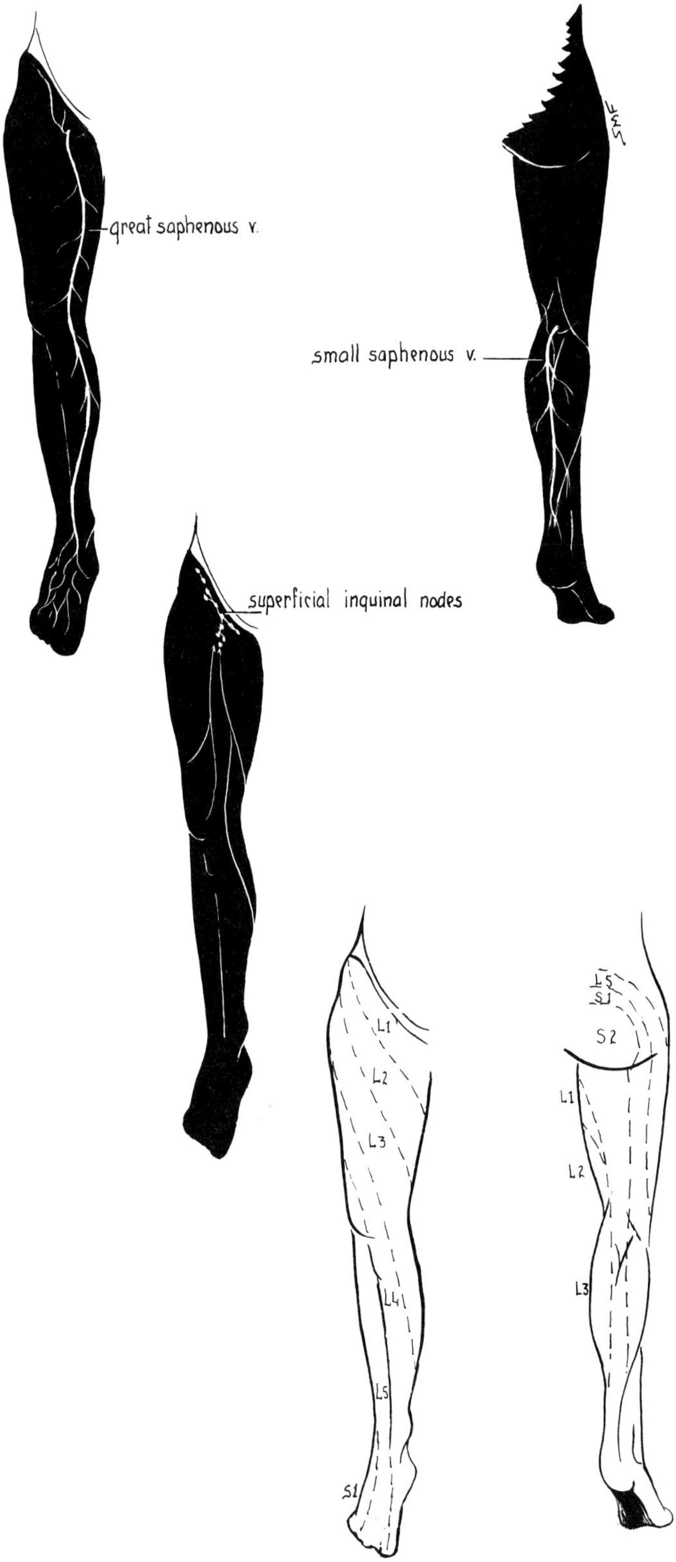
great saphenous v.
SMF
small saphenous v.
superficial inquinal nodes
L1
L2
L3
L4
L5
S1
L5
S1
S2
L1
L2
L3

The Back

The Spinal Cord (Spinal Medulla) and Meninges

The **spinal cord (spinal medulla),** a continuation of the medulla oblongata, lies within the vertebral canal of the vertebral column and extends down as far as L2. The present illustration depicts the general features of a segment of the cord as seen obliquely from behind. The nerve roots are shown only on the right.

The cord is cylindrical in shape with an H-shaped core of gray matter. The gray matter consists primarily of cellular elements, while the surrounding white matter consists largely of nerve fibers. Anteriorly, the *ventral median fissure* partly separates the cord into right and left halves while posteriorly, the thin *dorsal median septum* cuts into the cord.

The **dorsal root** (afferent fibers) of each spinal nerve begins as a linear series of rootlets. These rootlets emerge from that part of the cord just overlying the dorsal horn of gray matter. Similarly, the **ventral root** (efferent fibers) of the spinal nerve arises from the anterior aspect of the cord by a series of rootlets which begin right in the ventral horn of gray matter. These are not linear, but occupy a band 2 or 3 mm wide. Ventral and dorsal roots pass laterally, joining, usually within the intervertebral foramen, to form the spinal nerve (Vol. II, Fig. 9). The *spinal (dorsal root) ganglion,* which contains the cell bodies of the afferent fibers, lies on the dorsal root just proximal to this point of union.

Closely investing the cord throughout its length is the **pia mater.** This membrane provides sheaths for the vessels entering into the substance of the cord as well as for the nerve roots leaving it. The pial sheaths for the ventral and dorsal nerve roots extend right to their point of union. A *denticulate ligament,* of pial origin, extends longitudinally on either side of the cord. The apices of its toothed projections attach through the arachnoid to the dura.

Outside the pia, the cord is ensheathed by the thin-walled **arachnoid mater.** The arachnoid also sends a tubular sheath along the ventral and dorsal nerve roots for a short distance, but usually this sheath does not go as far as the pial sheath. Cobwebby fibers incompletely interrupt the space between the arachnoid and the pia. This *subarachnoid space* is filled with cerebrospinal fluid. Obviously, the subarachnoid space extends for a short distance on to the nerve roots.

The dense fibrous **dura mater** forms a tough outer encasement surrounding the arachnoid. A potential *subdural space* exists between the dura and the arachnoid. In general, the encasement for the emergent nerve roots which the dura provides is prolonged on to the spinal nerve. In the case of the very oblique lowest nerves, however, ventral and dorsal roots unite before reaching the dura so that only a single dural sheath is provided for the spinal nerve itself.

Between the bony walls of the spinal canal and the dura is the *epidural space.* The bony wall bounding the vertebral canal was seen in Volume II, Figure 1. Between the vertebrae, the canal is closed posteriorly by the ligamenta flava joining the laminae and anteriorly by the posterior longitudinal ligament joining the vertebral bodies.

Spinal nerves can be anesthetized in several locations. An anesthetic solution can be injected at the point where they emerge from the intervertebral foramina (paravertebral block), or into the epidural space (epidural and caudal anesthesia), or into the subarachnoid space (spinal anesthesia). The subdural space is less easily used.

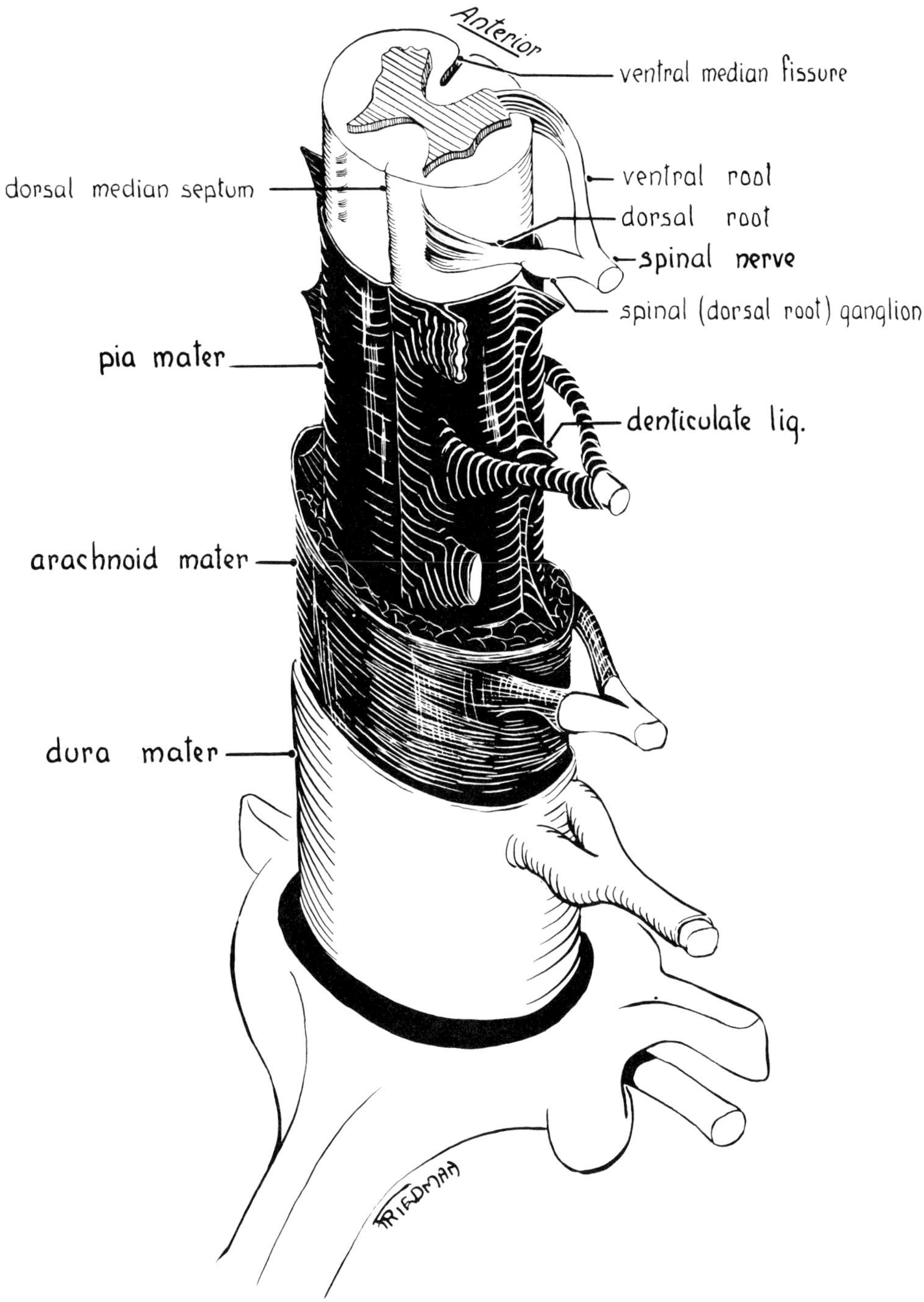
Anterior
ventral median fissure
dorsal median septum
ventral root
dorsal root
spinal nerve
spinal (dorsal root) ganglion
pia mater
denticulate lig.
arachnoid mater
dura mater
FRIEDMAN

The Spinal Cord

Two smooth fusiform enlargements are present on the cylindrical spinal cord. The *cervical enlargement* begins approximately opposite C2 and ends opposite T2, its greatest circumference being at the level of the roots of the 6th cervical nerve. The *lumbar enlargement* begins at approximately T9, attains a maximum at approximately T12, and tapers rapidly to end opposite L2 where the cord terminates.

The cervical and lumbar enlargements represent the areas where a large number of fibers concerned with the innervation of the upper and lower limbs arise. In the region of these enlargements the cord almost completely fills the vertebral canal so that a minimal bony displacement resulting from trauma in this area can cause injury to the cord.

The pia mater continues down as a thread beyond the termination of the cord at L2 to end by fusion with the dorsum of the coccyx. This *filum terminale* contains some nerve tissue within its upper part. The dura mater and the arachnoid mater end blindly opposite S2. Thus, between L2, where the cord terminates, and S2, the subarachnoid space is obviously very large.

The vertebral canal is continuous with the sacral canal which is closed by membrane close to the tip of the sacrum. Since the dura ends at S2, it is obvious that the epidural space between S2 and S5 is very large.

Since the spinal cord is shorter than the vertebral canal, it is obvious that in order to reach the intervertebral foramina the nerves must pass not only laterally but downward as well. Obviously, too, the first cervical nerve passes almost directly laterally to the intervertebral foramen, while the lumbar and sacral nerves pass almost vertically downward. Below L2, the subarachnoid space is occupied by those nerves which must still descend to reach their point of emergence. These nerves (L3 to coccygeal 1) are collectively referred to as the *cauda equina.*

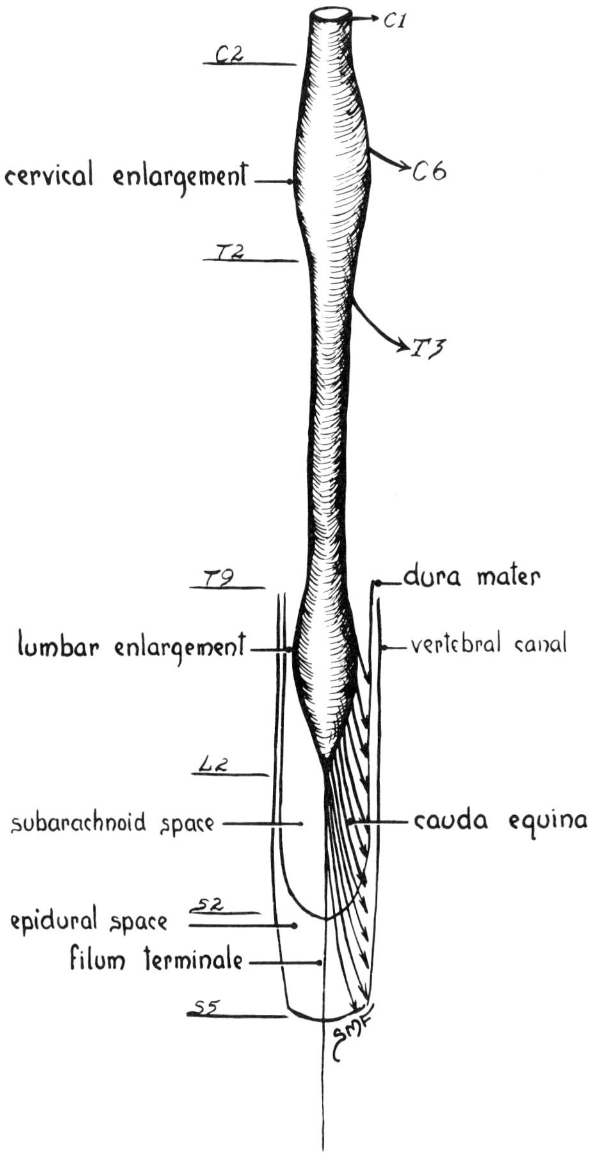
C1
C2
cervical enlargement
C6
T2
T3
T9
dura mater
lumbar enlargement
vertebral canal
L2
subarachnoid space
cauda equina
S2
epidural space
filum terminale
S5
SMF

The Framework

In this illustration, the lumbar vertebrae and sacrum are shown as they appear from behind and slightly obliquely. Of particular interest in this view of the **lumbar vertebrae** are the broad blunt *spines of the lumbar vertebrae.* These are readily palpated in the living subject since they lie immediately beneath the skin. In contrast to the thoracic region where the spines and laminae overlap one another, preventing any easy access into the vertebral canal, the spines here are almost a finger's breadth apart, and the laminae fail to meet by about the same amount. Only the ligamenta flava prevent access to the spinal canal between the laminae; consequently, needle puncture right into the subarachnoid space is readily effected in the lumbar region. This is not particularly desirable above L2 where the spinal cord might be damaged by an entering needle, but below this level such a lumbar puncture would encounter only the cauda equina, the constituent nerves of which tend to slide aside from the needle.

The line joining the highest points of the iliac crests across the back passes between the spines of the 3rd and 4th lumbar vertebrae or across the 4th lumbar spine. Since this is the most accessible and reliable landmark in the area, lumbar puncture is usually performed through the ligamenta flava joining either L3 to L4 or L4 to L5.

As previously mentioned, the **sacrum** is formed of five fused vertebrae. The *spines* of the upper four of these are usually present as vestigial knobs, while lateral to these the *laminae* have fused completely, closing off the vertebral canal, which is here termed the sacral canal. The *sacral canal* is occupied by those constituents of the cauda equina which have not yet reached their point of emergence as well as by the filum terminale.

The lower end of the sacral canal is open as the *sacral hiatus* which is flanked on either side by the readily palpated *sacral cornua.* In life the sacral hiatus is sealed by a dense ligament closing off the sacral canal. For purposes of anesthesia the sacral canal can be entered by piercing this ligament. Provided the needle is kept below S2, it will lie in the epidural space. This variety of epidural anesthesia is termed "caudal block." Because of the frequency with which such a block is performed it is important to recognize that anomalies of the sacrum are common. Among these may be mentioned the sacrum formed of 6 segments with a lengthened sacral canal, and that formed of 4 segments with a short sacral canal. In addition, fusion of spines and laminae may be incomplete, especially at the lower end, so that the sacral hiatus may be considerably enlarged.

Anomalies in the lumbar vertebrae are also common and may, in some cases, constitute adequate causes for strain and low back pain.

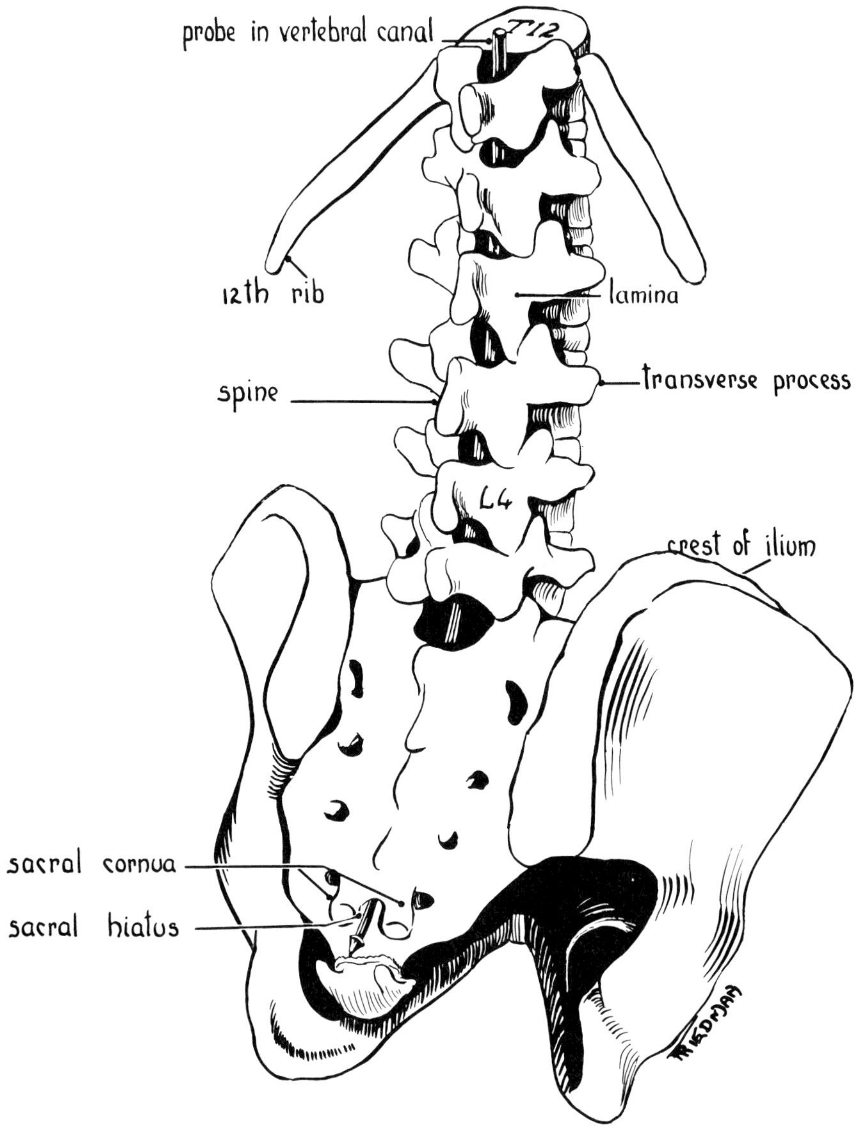
probe in vertebral canal
T12
12th rib
lamina
spine
transverse process
L4
crest of ilium
sacral cornua
sacral hiatus

The Muscles

The muscles of the back are numerous and complicated in their attachments. Clinically, only their general arrangement is of interest.

The muscle mass runs vertically on either side of the midline from the sacrum to the skull and, as a whole, may conveniently be termed the **"Extensor of the back."** It consists of a deep stratum lying close to the spinal column with fibers passing, in general, rather upward and medially, and a larger superficial stratum whose fibers are, in general, more vertical.

The parts of the deep stratum (left side of this diagram) are termed *Multifidus* (which overlies *Rotatores*) in the lumbar region and *Semispinalis* in the thorax and neck.

The more superficial stratum (right side of this diagram) is broader in its width than is the deep stratum. In the lumbar region this mass, termed **Erector Spinae,** arises from the sacrum, ilium, and lumbar vertebrae and passes upward toward the thorax where it separates into three columns. The most medial column, termed *Spinalis,* attaches chiefly to the thoracic spines. Lateral to Spinalis, the column is termed *Longissimus* (i.e., Longissimus Thoracis, Longissimus Cervicis, and Longissimus Capitis). The most lateral column is made up of the *Iliocostalis* group which is also named according to the region of its origin and insertion (e.g., Iliocostalis Lumborum, Iliocostalis Thoracis, and Iliocostalis Cervicis).

In general, the component parts of the Extensor of the back are supplied by the dorsal rami of adjacent spinal nerves.

As a whole, it should be appreciated that the extensor column of muscle is thickest in the lumbar region where it supports the lumbar curve of the spine, and again in the neck where it supports the cervical curve. The curves of the thorax and sacrum are maintained largely by the shapes of the articulating bones. These latter curves are termed "primary" since they are present in the fetus, while those of the lumbar and cervical regions, which develop postnatally, are termed "secondary" or "acquired."

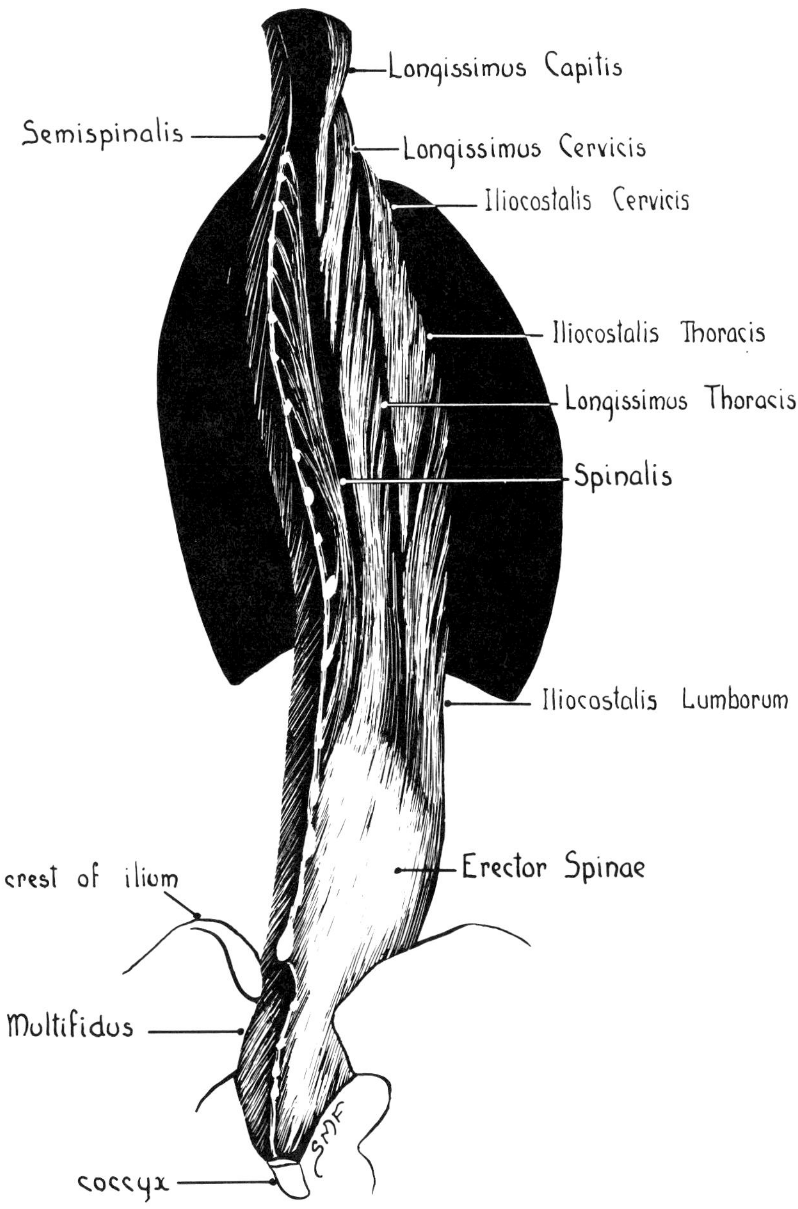
Longissimus Capitis
Semispinalis
Longissimus Cervicis
Iliocostalis Cervicis
Iliocostalis Thoracis
Longissimus Thoracis
Spinalis
Iliocostalis Lumborum
Erector Spinae
crest of ilium
Multifidus
coccyx

INDEX

(References are to figure numbers)